REINFORCED THERMOPLASTICS

REINFORCED THERMOPLASTICS

W. V. TITOW
Project Manager,
Yarsley Research Laboratories Ltd
(A Division of Fulmer Research Institute Ltd), Ashtead, Surrey, England

and

B. J. LANHAM
Manager, Thermoplastic Operations,
LNP Plastics, Nederland B.V., Breda, The Netherlands

A HALSTED PRESS BOOK

JOHN WILEY & SONS
NEW YORK—TORONTO

PUBLISHED IN THE U.S.A. AND CANADA BY
HALSTED PRESS
A DIVISION OF JOHN WILEY & SONS, INC., NEW YORK

Library of Congress Cataloging in Publication Data

Titow, W. V.
Reinforced thermoplastics.

"A Halsted Press book."
Includes index.
1. Thermoplastics. 2. Reinforced plastics.
I. Lanham, B. J., joint author. II. Title.
TP1180.T5T52 668.4'23 75–16335
ISBN 0–470–87518–6

WITH 67 TABLES AND 96 ILLUSTRATIONS

Filmset by Typesetting Services Ltd, Glasgow
Printed in Great Britain by Galliard (Printers) Ltd, Great Yarmouth

PREFACE

This book is an attempt to present a reasonably unified account of the technology and properties of reinforced thermoplastics, with some reference to a few commercial aspects. To the best of our knowledge no such text has been available so far, although the development of this comparatively new class of plastics materials has been vigorous, and the literature it has generated includes many good technical papers as well as a considerable volume of data from producers. We have referred to both these general sources, and both are represented in the bibliography sections of those chapters to which the references selected are relevant.

The theme and orientation of the book are essentially technological. We have centred attention, and the technical information content, on the nature and properties of the industrially important reinforced thermoplastics and their main constituents (the base polymers and reinforcing fillers), their methods of production and commercial sources, their processing, applications, and their role as engineering plastics. In keeping with the technological bias we have not attempted to deal in depth with what might be termed the scientific aspect of reinforced thermoplastics, that is to say the fundamentals of the way in which their properties are governed by their composition and fine structure. We have, however, devoted a section (Chapter 1, Section 1.4) to an outline of the main current ideas on, and understanding of, this aspect, and have referred to them as and where appropriate in other places. Numerical data on the properties of reinforced thermoplastics and their constituents are strongly featured throughout the text. We make no apology for the fact that the data are not expressed in SI units. A coherent system of units is undoubtedly useful to the scientist, but it is a fact of industrial life that technologists working with reinforced thermoplastics still talk (and largely write) in terms of the Imperial unit system in the Anglo-Saxon world and the technical metric system in Europe. The position may perhaps change in a few years time, but currently these are the units in general use in this particular field and it would therefore be illogical not to use them in a book of this kind. Conversion tables are provided in Appendix 3.

This book is the result of teamwork. Our warmest thanks go to the third member of the team, our friend Mrs. Rene Chizlett, without whose secretarial contribution the manuscript would never have come into being in the first place, nor finally gone into print.

W.V.T.
B.J.L.

CONTENTS

ACKNOWLEDGEMENTS

It is a pleasure to record our appreciation of the help we have received from a number of friends expert in some of the subjects discussed in the book. The help took various forms, from valuable comments and advice to provision of information and material. In this connection we are most grateful to two of our colleagues at the Yarsley Laboratories: Mr. D. K. Mugridge, who provided much of the information on moulding temperatures for the tables of Section 3.3 of Chapter 3, and Mr. J. A. Mead, for useful discussions of some physical properties of materials and for permission to use certain office facilities under his jurisdiction.

Professor M. Braden of the London Hospital Medical College, University of London, very kindly read the script of Chapter 6: his comments have been most helpful. Our thanks for items of information and/or illustrations go to Mr. F. Mandy of Owens-Corning Fiberglas Europe S.A. (Chapter 1, Section 1.2 and Chapter 8, Section 8.1), Messrs. F. D. Monk and D. Pickthall of Fibreglass Ltd. (Chapter 4, Section 4.2), Mr. R. C. Hopkins of Plastichem Ltd. (Chapter 4, Section 4.3), Mr. L. F. Parks of Cape Asbestos Fibres Ltd. (Chapter 4, Section 4.4), Messrs. D. A. Dryburgh and M. R. Ross-Gower of D. Dryburgh & Co. Ltd., Mr. J. D. Wilkes of Baker Perkins Ltd. (Chapter 5, Section 5.2.2.1.2) and Mr. R. J. Fowler of the Churchill Instrument Co. Ltd. (Chapter 7, Section 7.2.2.3). Sections 4.2, 4.3 and 4.4 benefited from the helpful and constructive comments made respectively by Messrs. Monk, Hopkins and Parks who were kind enough to read the original drafts.

We are much indebted to Mr. F. Burford of Kingston Polytechnic for taking electron-scan micrographs of reinforced-compound specimens (Figs. 5.2 and 5.3 in Chapter 5) and to Mr. S. Miyaki of the Asahi Glass Co. Ltd. who kindly provided the data on which Table 2.5 in Chapter 2 is based.

Permissions from companies to use certain data from their technical literature, and those from copyright holders to reproduce material from their publications, are acknowledged individually in the appropriate places in the text.

The frequent references we have made to the work and data of Mr. J. E. Theberge are a measure of our appreciation of the contribution he and his co-workers on the technical staff of the Liquid Nitrogen Processing Corporation have made to the technology of reinforced thermoplastics.

Finally, we would like to thank our respective employers for their approval of this undertaking.

W.V.T.
B.J.L.

CHAPTER 1

INTRODUCTION

1.1 REINFORCED THERMOPLASTICS: THEIR ORIGIN, WHAT THEY ARE AND WHY AND WHERE THEY ARE USED

The modern reinforced thermoplastics consist essentially of the thermoplastic polymer (the 'base polymer') in which the reinforcing filler has been dispersed. By far the most widely used reinforcing fillers are fibres, predominantly glass and asbestos fibres. Minor additives may also be present.

Before discussing the general nature of reinforced thermoplastics it may be useful to consider the way in which their parent thermoplastics attained their current importance, and to mention the advantages which led to their outstripping their forerunners, the thermosetting plastics, in so many applications.

Thermosetting compounds, in particular phenolic moulding materials, and their fabrication by compression moulding, became commercially important in the 1920s; urea and melamine moulding materials followed in the 1930s. Practically all thermosetting moulding compounds consisted predominantly of a resin and a filler. In the phenolics the filler was normally woodflour, mica or glass: in the aminoplastics it was woodflour or cellulose. These fillers were necessary to impart increased strength and stability, and useful moulding compositions were not possible without them.

The major thermoplastics of the pre-war years were the cellulosic plastics and in particular cellulose acetate. These tough horny materials did not need the inclusion of fillers to be useful as moulding compounds. In fact they were often employed in applications where their transparency was important, and this would have been adversely affected by fillers. Following World War II polystyrene took over from cellulose acetate as the work-horse of the plastics moulding industry. This was a material that moulded easily and could be satisfactorily employed in a wide range of applications without the inclusion of fillers. Polystyrene and the other thermoplastics were processed by the injection moulding technique which at that time could not be used on thermosetting plastics. This method of fabrication offered a much greater degree of automation than was normally achieved with compression mouldings and the rate of production was also higher. This major processing advantage coupled with the toughness and flexibility of the recently introduced polyethylene and the low cost and versatility of polystyrene and high impact polystyrene caused the volume of thermoplastics being converted to surge ahead of that of the thermosetting materials.

In 1951 an American, R. Bradt, was faced with the need to provide a material with closely specified properties for a military application—a mine case. He found that the material specification could be met by the inclusion of glass fibres in polystyrene. The glass-reinforced material he provided was produced by a coaxial extrusion technique similar to that used in wire covering, i.e. continuous glass rovings were passed through a cross-head die and extrusion-coated with the polymer. The strand formed was cut to $\frac{1}{2}$ or $\frac{1}{4}$ in (12·7 or 6·35 mm) long granules for moulding. Such products offered almost twice the tensile strength of unreinforced polystyrene and were four times as rigid. The company formed to manufacture these materials, Fiberfil Inc., was the first supplier of reinforced thermoplastics.

At about the same time nylon 6.6 was being used as a moulding material in the USA for mechanically functional rather than decorative applications; it was the first of the so-called engineering plastics. It offered good toughness coupled with high wear resistance and a low coefficient of friction. This combination was ideal in applications like bushes, gears and sliders of all types. Its main disadvantages were poor dimensional stability and comparatively low rigidity and strength. These features could be improved with glass reinforcement. Fiberfil, using the Bradt technique, started making and marketing a glass-reinforced nylon which rapidly became its most important product.

In the early 1960s ICI in the UK, Bayer in Germany and the LNP Corporation in the United States developed new grades of glass-reinforced nylon. These were produced by a different, basically older and more conventional technique, *viz.* extrusion compounding. Some early work of DuPont, as illustrated e.g. by the Specification of their 1946 UK Patent (618 094), had shown that the inclusion of glass fibre into nylon by extrusion compounding resulted in a useful improvement in physical properties. Refinement of this process by ICI and Bayer resulted in the commercial availability of the first grades of so-called 'short glass fibre' products. This term was used to distinguish these materials from the Fiberfil-type products wherein the fibres were completely aligned along the granule and of the same length as the granule ($\frac{1}{4}$ in (6·35 mm)). The extrusion compounded products could have mean fibre lengths as low as 0·010 in (0·25 mm). They did not offer strength properties on a par with products reinforced with 'long' glass fibres, but were sufficiently good to meet the requirements of most of the applications for which the latter were used. Moreover, they offered significant advantages in terms of surface finish and processability. Once these products were accepted in the USA and in Europe the range of polymers available in reinforced form increased, and their sources multiplied rapidly.

In the USA the use of the main commodity plastics, polystyrene and polyethylene, and the quasi-engineering plastics, acrylonitrile/butadiene/styrene (ABS), styrene/acrylonitrile (SAN) and polypropylene in reinforced form grew rapidly in importance due to their utilisation in the automotive industry. In Europe these materials did not take on the same importance but nylon in reinforced form was widely exploited and, on a much smaller volume level, polycarbonate became the next most useful reinforced thermoplastic.

The following section deals with the statistics of the growth of reinforced thermoplastics in the USA, Europe and Japan.

1.2 STATISTICS

At the present time the only reliable figures for reinforced thermoplastics consumption are those published for the USA. However various companies active in the field have put forward figures for Europe. Figures for Japan are even more difficult to obtain. Japanese reinforced thermoplastics are not exported to the West in any great quantities although some are well known, e.g. the glass-reinforced polyester Teijin's FR-PET—*cf.* Chapter 4. Such data as are available are not complete enough to warrant their inclusion in this section on a par with the better-documented Western information.

The growth of reinforced thermoplastic materials in the USA is shown in Table 1.1. It can be seen that with the exception of the economic recession year of 1970 the materials have been growing at the rate of over 20% per annum. This has resulted in the total consumption practically doubling in five years. It is also interesting to note the advance of the polyolefins to become the largest tonnage group. Polycarbonate, modified polyphenylene oxide (PPO) and polysulphone make up the 'polyarylates' and it is evident that this group is rapidly overhauling the nylon-based materials.

The percentage break-down of the usage pattern in the USA for the early 1970s is shown in Table 1.2. The automotive industry is of prime importance, accounting for over 50% of the total reinforced thermoplastic materials used. Domestic appliances and business machines rank second.

The statistics for Europe are sparse; Tables 1.3 and 1.4 below show partly estimated figures for 1970 with projection for 1975 and 1980. Figures 1.2, 1.3, 1.4 and 1.5 are consumption graphs from three different sources.

TABLE 1.1
USA Reinforced Thermoplastics Market by Polymer Type (*tonnes*)

Polymer type	1968	1969	1970	1971	1972	1973[a]
Nylons	5 450	6 350	5 900	6 350	7 250	8 620
Polyolefins	7 250	8 200	8 200	10 500	14 500	16 810
Styrenics	5 900	8 650	10 000	11 000	11 350	12 750
Polyarylates	1 800	2 050	2 050	4 100	6 350	8 620
Acetal and others	2 050	1 800	910	1 800	3 650	7 710
Total	22 450	27 050	27 060	33 750	43 100	54 510
Growth %		20	0	24	28	38

[a] The percentage distribution of glass fibre reinforced thermoplastics for 1973 is shown in Fig. 1.1 (F. Mandy, paper presented at the Semaine des Injecteurs 1974, Brussels, Belgium: reproduced with permission of the author).

TABLE 1.2
USA Reinforced Thermoplastics Market by Application

Application area	Percentage
Automotive	56
Domestic appliances and business machines	27
Military	4
Electrical and engineering	4
Others	9

TABLE 1.3
European Reinforced Thermoplastics Market by Polymer Type (*tonnes*)
(Courtesy of Mr. F. Mandy and Owens-Corning Fiberglas Europe S.A.)

Polymer type	1970	1975	1980
Nylons	15 100	47 200	82 500
Polyolefins	1 710	9 000	31 500
Styrenics	428	4 500	13 500
Polyarylates	3 000	10 500	18 000
Acetal and others	1 070	3 750	4 500
Total	21 308	74 950	150 000

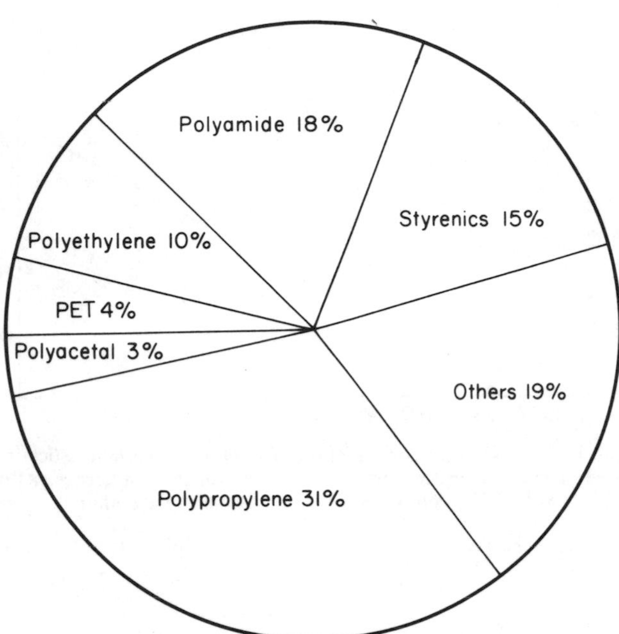

FIG. 1.1 Percentage breakdown: USA usage of glass-fibre-reinforced thermoplastics by polymer type in 1973. (Courtesy of Mr. F. Mandy and Owens-Corning Fiberglas Europe S.A.)

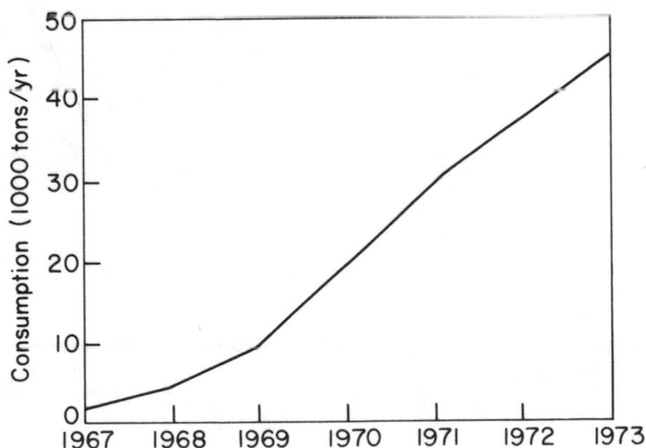

FIG. 1.2 Consumption of reinforced thermoplastic materials in Europe. (Reprint by permission of *Modern Plastics International*, McGraw-Hill Inc.)

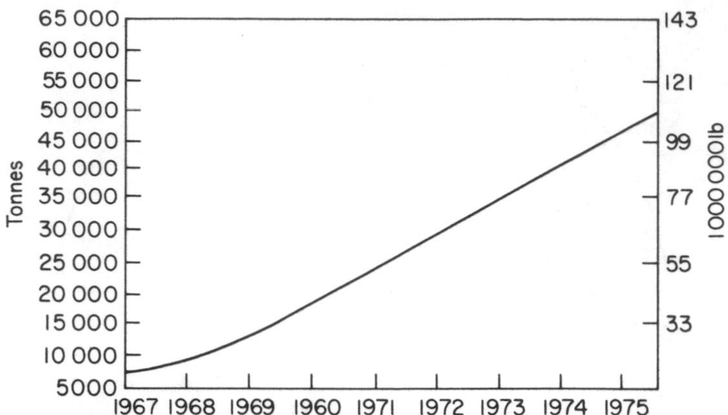

FIG. 1.3 Another estimation of consumption of reinforced thermoplastics in Europe in recent years. (L. Lesseliers, Proceedings of the 27th Annual Conference, RP/C Institute, SPI, 1972: reproduced with permission of the author.)

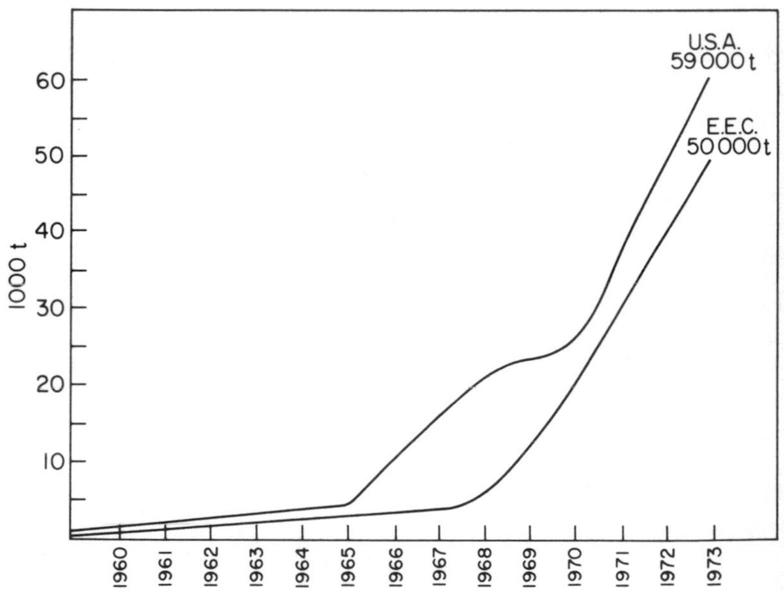

FIG. 1.4 Growth of glass-fibre-reinforced thermoplastics in the USA and Europe. (Courtesy of Mr. F. Mandy and Owens-Corning Fiberglas Europe S.A.)

TABLE 1.4
European Reinforced Thermoplastics Market by Application Area (*tonnes*)
(Courtesy of Mr. F. Mandy and Owens-Corning Fiberglas Europe S.A.)

Application area	1970	1975	1980
Automotive	6 000	26 300	68 500
Business machines and appliances	8 750	25 500	40 500
Electric/electronic	4 500	14 300	24 000
Other	2 058	8 850	17 000
Total	21 308	74 950	150 000

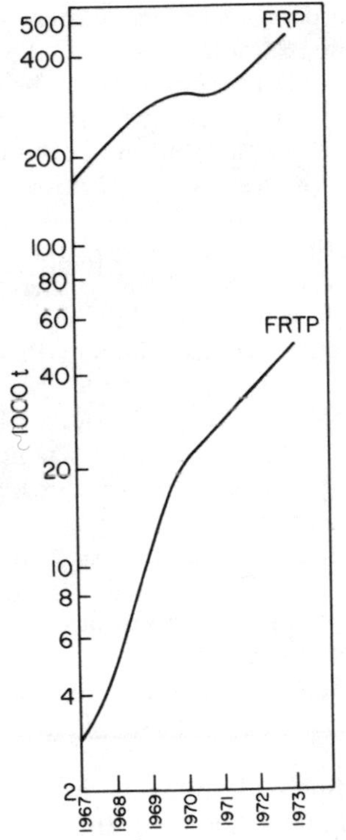

FIG. 1.5 EEC consumption of glass-fibre-reinforced plastics and glass-fibre-reinforced thermoplastics. (Courtesy of Mr. F. Mandy and Owens-Corning Fiberglas Europe S.A.)

1.3 THE TECHNOLOGY AND USES OF REINFORCED THERMOPLASTICS—PLAN AND SCOPE OF THE BOOK

Reinforced thermoplastics constitute a vigorous sector of the thermoplastics industry growing in importance both commercially and technically. An outline of their development and commercial status has been given in the preceding two sections. The reinforced thermoplastic materials available commercially and their main sources are reviewed in detail in Chapter 2. The remaining chapters deal with the technology of the main constituents (polymers and reinforcing fillers), the production, processing, applications and some special forms of reinforced thermoplastics. The book is thus essentially technological in its concept and treatment of the subject. This involves no inadvertent error of omission nor is it the intention to minimise the importance of the considerable amount of scientific work past and current directed to the elucidation of the fundamentals of the way in which the nature and form of the reinforcement influence the structure and properties of reinforced thermoplastics. However, as often happens in successful new technical developments, what might be called the technological art of these materials is still ahead of their science and the balance is unlikely to be redressed in the next few years. Enough well-founded, direct practical knowledge is available on the production, technical properties and applications of reinforced thermoplastics to justify the hope that its presentation in the form of a reference book will serve a useful purpose. By contrast there is as yet no unified fundamental theory of the structure and properties of reinforced thermoplastics. Probably the most comprehensive theoretical treatment of fibre-reinforced materials in general is that produced recently (1972) by Hashin[1] but even this does not deal fully with what it describes as 'the very difficult subject of physical strength'. Specifically with regard to fibre-reinforced thermoplastics, recent work by Abrahams and Dimmock[2] appears to bear out their claim that the strength and stiffness of such composites can now be predicted with some accuracy from the properties of the components coupled with suitable data on fibre distribution and orientation. These authors also claim good accuracy in property/cost predictions. Thus the considerable volume of available fundamental information, valuable as much of it undoubtedly is, still presents a rather fragmented picture containing areas which, if not actually blank, are certainly still rather diffuse. An attempt at a complete presentation of the theoretical aspects of reinforced thermoplastics would be premature at this stage because much that is fundamental to the subject and relevant to its full understanding still remains to be elucidated. Whilst the situation could change in the course of the next few years, at present it seems right in a book of this kind to focus attention on the technological aspect where solid, concrete information can be offered.

However, for the sake of completeness and to assist the interested reader, the main theoretical aspects of reinforced thermoplastics are very briefly outlined in the following section, whilst the sources mentioned should, it is hoped, provide a useful aid to those seeking a closer acquaintance with the subject.

1.4 SOME THEORETICAL ASPECTS OF THE STRUCTURE AND PROPERTIES OF REINFORCED THERMOPLASTICS

In a reinforced thermoplastic material the reinforcement filler, most commonly a fibre, is dispersed in a polymer matrix. In comparative terms the polymer (matrix) component is of low strength, fairly tough and extensible, of lower density and higher coefficient of thermal expansion than the reinforcing filler component which is strong, stiff and brittle. It is therefore the filler which should, ideally, ultimately bear as much as possible of any load or stress applied to the system, whilst the polymer matrix should effectively transmit the load to the filler. It is widely recognised that maximum benefit is derived from reinforcement when the conditions accord with this general principle.[3] It is also clear that the reinforcing filler will normally contribute heat resistance and dimensional stability to the system.

Whilst the general principles are well appreciated, a quantitative theoretical treatment of the relationships and mechanisms involved is made difficult by the complexity of the physical systems represented by reinforced thermoplastic materials. The treatments available so far relate mostly to certain basic cases or simplified systems.

The main factors which have to be taken into account and whose combined effects must be covered by any comprehensive theory are mentioned below. Even a simple listing gives some idea of the complexity of the problem. The difficulties are magnified still further by departure from ideal conditions within the individual factors (for example the fact that the polymeric matrix materials are not truly elastic but viscoelastic) and by the time and temperature dependence of matrix properties.

Fibrous fillers are the most common form of reinforcement in reinforced thermoplastic materials. From the theoretical standpoint a spherical filler particle or, more generally, a particulate filler with reasonable sphericity, may be regarded as the simplest case of a fibre with the aspect ratio (length-to-diameter ratio) of 1. With normal fibres which, by definition, are characterised by high aspect ratios, the theoretical analysis of the composite system becomes more complicated because fibre length and diameter become factors, as well as fibre orientation effects; these latter in particular can give rise to anisotropy of properties in the composite. For all these reasons, and in order to avoid gross over-simplification, the filler is assumed to be fibrous for the purpose of the present section.

The main effect of reinforcing filler is to increase the stiffness (modulus) and the strength of the polymer. The simplest theoretical treatment of these effects is based on a simple rule of mixtures.[4] The resulting approximate relationship, valid for a composite comprising continuous fibres uniaxially oriented within a uniform elastic matrix, tested in the direction of orientation, simply expresses the modulus or tensile strength of the composite as a sum of the products of the moduli or tensile strengths and volume fractions of the two components (the polymer matrix and reinforcement fibres),[4–6] i.e.

$$E_c = E_f V_f + E_m(1 - V_f) \tag{1}$$

$$T_c = T_f V_f + T_m(1 - V_f) \tag{2}$$

where E_c, E_m, E_f = moduli of, respectively, the composite, matrix polymer and reinforcing fibre; T_c, T_m, T_f = the analogous strengths; V_f = volume fraction of fibre; and $(1 - V_f) = V_m$, the volume fraction of matrix polymer.

The equations can be modified by introducing an alignment factor (η) which represents the combined effects of fibre misorientation (i.e. departure from uniaxial alignment) and any difference between the direction in which the material is tested and that of fibre orientation. Then

$$E_c = \eta E_f V_f + E_m(1 - V_f) \tag{3}$$

$$T_c = \eta T_f V_f + T_m(1 - V_f) \tag{4}$$

It may be noted that equations (1)–(4) do not take into account the critical volume of reinforcement which is the volume of the fibrous filler required to bear a load equal to that which could be sustained by the unfilled matrix.[7]

The approach based on the rule of mixtures has been extended to include the case of discontinuous reinforcement fibres of finite length.[5,8] With such fibres the load is transferred from the matrix to the fibre through the effect of shear at the polymer/fibre interface,[5,8–10] and the actual fibre length as well as interfacial adhesion between fibre and matrix become important parameters (see below).

Mathematical analysis of the properties of fibre-reinforced composites in terms of the main parameters has produced equations like that put forward by Chen[8] for a randomly-oriented, discontinuous-fibre composite:

$$\sigma_\theta = \frac{2\tau_m}{\pi}\left(2 + \ln \frac{\xi \sigma_c' \sigma_m}{\tau_m^2}\right) \tag{5}$$

where: σ_θ = strength of the randomly-oriented fibre composite; τ_m = shear strength of the material at the weakest interface; ξ = strength efficiency factor (=1 for continuous fibre composites); σ_c' = longitudinal strength of a continuous fibre composite; and σ_m = strength of the matrix material.

Theoretical approach to the prediction of the modulus and—to a lesser extent—the strength of fibre/thermoplastic composites has been developed to a considerable degree of refinement, with the aid of computer analysis, by Abrahams and Dimmock.[2] Their treatment, which is in part based on earlier theoretical work[11–19] includes the effects of the viscoelastic nature of the thermoplastic matrix and the time dependence of its properties (*cf.* also sub-section 1.4.3 below).

It is evident from the foregoing that the main factors which must be considered in any theoretical treatment of the structure and behaviour of reinforced thermoplastic materials will include:

1.4.1 Filler Content

This is normally expressed as a volume or weight fraction. Some effects in reinforced thermoplastics are simple functions of the volume fraction

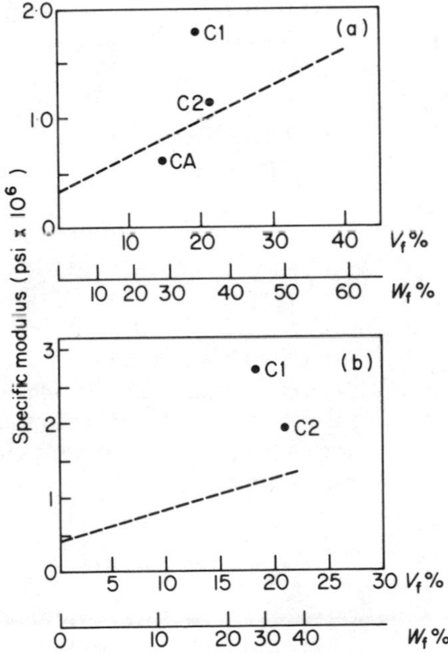

FIG. 1.6 Specific modulus as a function of volume and weight fractions of reinforcing carbon fibre in (a) nylon 6.6 and (b) styrene acrylonitrile copolymer.[3] (C1: carbon fibre type 1 RAE; C2: carbon fibre type 2 RAE; CA: carbon fibre Courtaulds.) (Reprinted, with permission, from Proceedings of the SPE RETEC, Toronto 1970.)

of the filler.[20] For example, specific modulus of styrene/acrylonitrile copolymer and nylon 6.6 reinforced with carbon fibres was found to be linear with the volume fraction, but not the weight percentage, of the filler,[3] *cf.* Fig. 1.6.

In general, other things being equal (and in particular given a good interfacial bond between the filler and the matrix), the filler content determines the extent to which the properties of the composite are modified from those of the matrix towards those of the filler. The trend is illustrated by equations (1)–(4).

1.4.2 Filler (fibre) Characteristics

1.4.2.1 Fibre Length and Diameter

As already mentioned, when the fibres are of finite length, stress is assumed to be transferred from the matrix to the fibre by a shear transfer mechanism. This gives rise to the concept of 'critical' fibre length (for a given fibre diameter), i.e. the shortest length necessary to enable the stress in the fibre to build up to the value of tensile fracture stress.[3] Thus at lengths below the critical, stress transfer to the fracture-stress level cannot occur. The fibres are not long enough to be effectively gripped by the matrix to 'take the strain' and break under tensile loading. Instead they will slip and be pulled out.[5]

The critical length (L_c) can be calculated from relationships of the type[5,6,9,21]

$$L_c = \frac{DT_f}{2A} \tag{6}$$

where: D = diameter of fibre; T_f = tensile strength of fibre (≡breaking stress); and A = strength of bond between fibre and matrix (often assumed equal to the strength of matrix polymer).

In practice a spectrum of fibre lengths may be present in reinforced thermoplastic composites. Bowyer and Bader[9] propose the following equation to express the effect of this, in conjunction with various other factors involved, on the stress in the composite (σ_c):

$$\sigma_c = C\left[\Sigma \frac{\tau L_i V_i}{2r} + \Sigma E_f e_c\left(1 - \frac{E_f e_c r}{2L_j \tau}\right) V_j\right] + E_m e_c(1 - V_f) \tag{7}$$

where: C = fibre orientation factor; τ = shear strength of fibre/matrix bond; r = fibre radius; L_i = fibre length for the 'subcritical' fibre fraction; V_i = volume fraction of subcritical fibres; L_j = fibre length for the 'supercritical' fibre fraction; V_j = volume fraction of supercritical fibres; V_f = total volume fraction of fibre; e_c = strain in composite; and E_f, E_m are as previously defined.

The same authors suggest that the optimum fibre length for glass and carbon fibre reinforcement is about 1–2 mm.[22]

Ample practical data are available on the effect of fibre length on tensile strength, modulus, impact strength, shear strength, mould shrinkage and other properties of various reinforced thermoplastic materials.[6,21–24]

1.4.2.2 Fibre Strength and Stiffness

The general effects of intrinsic fibre strength and modulus are illustrated by equations (1)–(7). An important point with regard to the difference in

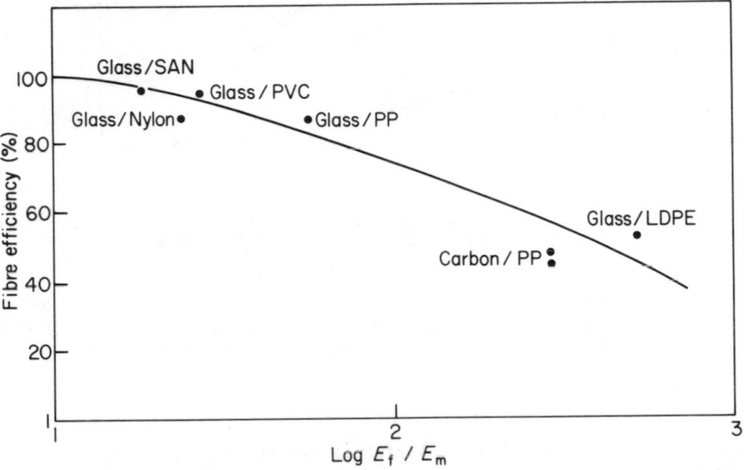

Fig. 1.7 Fall in fibre efficiency with fibre to matrix modulus ratio in short fibre composites. (Data of Davis[5]: courtesy of the Plastics Institute.)

modulus between the fibre and the matrix has been made by several workers:[5,25,26] this is that, as illustrated e.g. by Davis' data,[5] *cf.* Fig. 1.7, the efficiency of utilisation of the stiffness of reinforcement fibres decreases with increasing difference in modulus between fibre and matrix. Davis suggests 50:1 as the optimum fibre: matrix modulus ratio.[5]

Attention has been drawn by Thomas[27] to the effect of glass filament (continuous) diameter on the intrinsic strength of the filament and hence on the strength of filament strands embedded in polymers (*cf.* Fig. 1.8). This author advocates the use of glass fibres of about 0.005 mm for maximum strength of the composite.

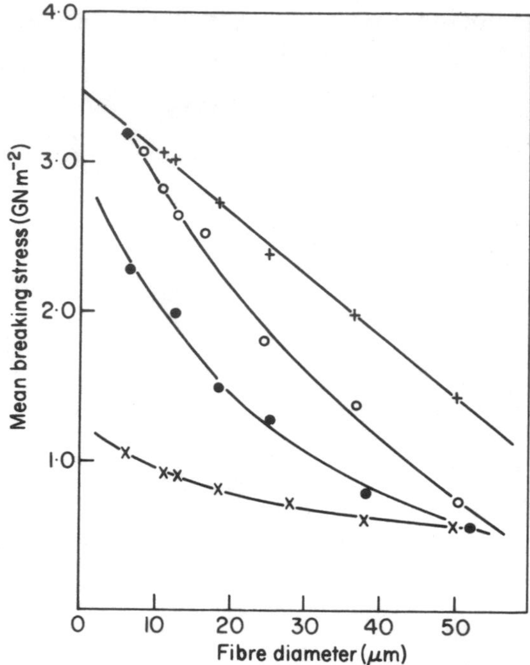

FIG. 1.8 Effect of fibre diameter on breaking stress in various resin matrices. × glass strand; ● strand in polystyrene; ○ strand in epoxide; + strand in polyester. (Data of Thomas[27]: reproduced with permission of the author and *Nature*, Macmillan Journals Ltd.)

1.4.3 Matrix Properties

Some general theoretical considerations relating to the contribution of the matrix polymer to the behaviour of reinforced thermoplastics have already been mentioned above. These considerations apply in conditions of low stress and short test or service times, and it is a most important general feature of the reinforced thermoplastic composites that the matrix polymers are viscoelastic, and that their behaviour in tests or service changes at high stress levels and/or over long times and/or at elevated temperature, giving rise to the phenomena of creep, stress relaxation and static or dynamic fatigue. The theory of these phenomena even in unfilled thermoplastic polymers is beyond the scope of the present section. It is discussed in many publications, e.g. those by Turner[28-31] and others.[32-34]

A considerable amount of experimental data has been accumulated on the long-term, high-stress and high-temperature behaviour of reinforced thermoplastics. From this it is clear that the effect of the matrix on the properties of the composite increases in those conditions, although conversely the presence of the filler will generally improve those properties in comparison with the matrix polymer alone.[2,5,6,23] It has been suggested that at glass fibre loadings above 40% by weight the time-dependent component of deformation of reinforced thermoplastics becomes negligible.[3] Prediction of the long-term behaviour of reinforced thermoplastics, based on theory and data from evaluation tests, is of obvious importance in forecasting the service performance of these materials. The subject has received a great deal of attention and progress has been made especially with the use for such predictions of creep and fatigue test data.[23,35–38] However, the theory is not yet complete, although—as mentioned above—some time-dependent properties of reinforced thermoplastic composites have been predicted in some cases from those of the constituents.[2]

Davis[5] gives the following list of conditions 'encouraging', i.e. bringing out in the composite, respectively the properties associated with the matrix, and those associated with fibrous reinforcement:

Conditions encouraging matrix properties	*Conditions encouraging fibre properties*
High strain conditions	Low strain conditions
Long time	Short time
High temperature	Low temperature
Poor bonding	Good bonding
Short fibres	Long fibres
High modulus ratio	Low modulus ratio

1.4.4 Filler/Matrix Interface

From the theoretical as well as the practical standpoint the most important feature of the interface is the adhesion of matrix material to the filler. As has been mentioned, good interfacial adhesion is necessary for effective transfer of stress. In its absence even long fibres will pull out of the matrix without breaking, which vitiates the reinforcement effect.[39] In a simplified theoretical treatment the adhesion effect may be represented by means of an 'efficiency factor' in expressions of the same type as equation (3), e.g.[3]

$$E_c = V_m E_m + k_{eff} V_f E_f \tag{8}$$

in which k_{eff} represents the degree of adhesion.

The practical importance of interfacial adhesion is amply illustrated by test data,[6,23,40] especially those relating to the improvement of both short- and long-term mechanical properties of the composite as a result of pre-treatment of the fibrous filler with 'coupling agents' to promote adhesion to the polymer in glass-reinforced polyolefines.[24,35,39,41,42] Most of the interfacial adhesion studies have been carried out in this area.[43–45]

Positive 'chemical' bonding (via covalent linkages) at the interface is believed to be the most effective bonding mechanism[46,47] in interfacial adhesion. Chemical compatibility permitting intimate contact and operation of secondary forces is also a factor. Other factors include the microconfiguration and frictional properties of the filler surface and the purely geometrical effect of the area of contact which is a function of the specific surface (and hence the diameter and aspect ratio) of the filler.[5]

REFERENCES

1. Hashin, Z. (1972). NASA Contractor Report CR-1974.
2. Abrahams, M. and Dimmock, J. (1971). *Plastics and Polymers,* **39,** 141, 187.
3. Maine, F. W., Riseborough, B. E. and Theberge, J. E. (1970). SPE RETEC., 'Polymer Structures and Properties', Toronto.
4. Kelly, A. (1973). *Strong Solids,* Oxford University Press.
5. Davis, J. H. (1971). *Plastics and Polymers,* **39,** 140, 137.
6. Williams, J. C. C., Wood, D. W., Bodycot, I. F. and Epstein, B. N. (1968). 23rd ANTEC of SPE Reinforced Plastics/Composites Division, Proceedings Section, 2C, 1.
7. Kelly, A. and Davies, G. J. (1965). *Metallurgical Reviews,* **10,** 37.
8. Chen, P. E. (1971). *Polymer Eng. and Sci.,* **11,** 1, 51.
9. Bowyer, W. H. and Bader, M. G. (1972). *J. Mat. Sci.,* **7,** 1315.
10. Kelly, A. and Tyson, W. R. (1965). *J. Mech. Phys. Solids,* **13,** 329.
11. Halpin, J. C. (1969). *J. Comp. Materials,* **3,** 732.
12. Ashton, J. E., Halpin, J. C. and Petit, P. H. (1969). *Primer on Composite Materials: Analysis,* Technomic Publishing Co., Stamford, Conn., USA, Ch. 5.
13. Adams, D. F. and Doner, D. R. (1967). *J. Comp. Materials,* **1,** 1, 4.
14. Adams, D. F. and Doner, D. R. (1967). *J. Comp. Materials,* **1,** 2, 152.
15. Tsai, S. W. (1966). AFML-TR-66-149, Part II.
16. Halpin, J. C. and Pagano, N. J. (1969). *J. Comp. Materials,* **3,** 10, 720.
17. Kelly, A. (1964). *Proc. Roy. Soc.* A **282,** 63.
18. Azzi, V. D. and Tsai, S. W. (1965). *Experimental Mechanics,* **9,** 283.
19. Lees, J. K. (1968). *Polymer Eng. and Sci.,* **8,** 3, 195.
20. Loveless, H. S. and McWilliams, D. E. (1970). *Polymer Eng. and Sci.,* **10,** 3, 139.
21. Filbert, W. C., Jr. (1969). *SPE Journal,* **24,** 1, 65.
22. Bader, M. G. and Bowyer, W. H. (1973). *Composites,* **4,** 4, 150–156.
23. Crabtree, J. D. and Pickthall, D. (1969). Paper at I.R.I. Conference, Loughborough, 15–17th September.
24. Roos, G. (1970). *Kunststoffe,* **60,** 12, 924.
25. Anderson, R. M. and Lavengood, R. E. (1968). *SPE Journal,* **24,** 3, 20.
26. Hollingsworth, B. L. and Sims, D. (1969). *Composites,* **1,** 2, 80.

27. Thomas, W. F. (1973). *Nature,* **242,** 5398, 455.
28. Turner, S. (1973) in *The Physics of Glassy Polymers* (Ed. R. N. Haward), Applied Science Publishers, London.
29. Turner, S. (1969). 'Engineering Design with Polyolefins', 27th ANTEC of the SPE, Chicago, Ill., May 5–8.
30. Turner, S. (1966). *Plastics Institute Transactions and Journal,* **34,** 127–135.
31. Turner, S. (1969). 'Deformation Data for Engineering Design', *Testing of Polymers* (Ed. W. E. Brown), Vol. 4, Interscience, New York.
32. Ratcliffe, W. F. (1966). *Plastics Institute Transactions and Journal,* **34,** 137.
33. Morris, A. C. and Gill, R. A. (1967). *Plastics,* **32,** 360, 1250.
34. Powell, P. C. (1969). *Progressive Plastics,* **9,** 73.
35. Cessna, L. C., Jr. (1970). 28th ANTEC of the SPE, Section 25, p. 527.
36. Pohrt, J. (1973). 'A Critical Strain Method for the Rapid Determination of Long-term Cracking in Thermoplastics and its Dependence on Internal Structure', Paper at Symposium at the Polytechnic of the South Bank, London.
37. Ogorkiewicz, R. M. and Turner, S. (1971). *Plastics and Polymers,* **39,** 141, 209.
38. Gotham, K. V. (1972). *Plastics and Polymers,* April, 59.
39. Cessna, L. C., Thomson, J. B. and Hanna, R. D. (1969). *SPE Journal,* **25,** 35, Oct.
40. Wicker, G. L. (1973). 'Asbestos Reinforced Thermoplastics', Paper at Symposium on Advances in Reinforced Thermoplastics, Kingston Polytechnic, 13th April.
41. Anon. (1968). *Plastics Technology,* **14,** 10, 23.
42. Hartlein, R. C. (1970). *Ind. Eng. Chem. Prod. Res. Develop.,* **10,** 1, 92.
43. Erickson, P. W. (1970). 25th Conference, RP/C Division, SPI, Paper 13-A.
44. Oswitch, S. and Golownia, R. F. (1970). *Reinf. Plast.,* **5,** 525.
45. Sterman, S. and Marsden, J. G. (1966). American Ceramic Society Meeting, Glass Division Symposium, Washington.
46. Plueddemann, E. P. (1965). 20th Conference, SPI, Paper 19-A.
47. Plueddemann, E. P. (1972). 27th Conference, RP/C Institute, SPI, Paper 11-B.

CHAPTER 2

COMMERCIAL REINFORCED THERMOPLASTICS

All the major suppliers of reinforced thermoplastics (and many minor ones) are located in Europe, the USA and Japan. This chapter provides information on commercial reinforced thermoplastic materials and their sources.

The main producers and their products are listed separately for the three major areas (Tables 2.1, 2.3 and 2.5). In addition, for Europe (including the UK) and for the USA, the products offered by each manufacturer are described by reference to their trade name, polymer type and grade, and type of reinforcement. Where it has been practicable, information on the level of reinforcement and on other additives present has also been included.

Broadly, producers of reinforced thermoplastics fall into two categories:

(1) major polymer manufacturers, most of whom offer reinforced versions of some of their products; and
(2) independent manufacturers, who are not in basic polymer production, but purchase polymers from the primary producers and combine them with suitable reinforcements.

In the United States the independent producers were historically the first suppliers of reinforced thermoplastics, whereas in Europe the polymer manufacturers were the first to offer reinforced grades. In the latter area it is only recently that independent manufacturers have become commercially important. In Japan the major suppliers are the polymer manufacturers.

2.1 EUROPE AND THE UK

The major manufacturers of reinforced thermoplastic materials in Europe, including the UK, are listed in Table 2.1; the list includes the polymers they provide in reinforced form. Table 2.2 contains details of grades of reinforced thermoplastic materials produced and marketed in Europe. It will be noticed that not all of the companies shown in Table 2.1 have their products listed in Table 2.2. Those companies whose products have been omitted are either subsidiaries or licensees of American companies and

TABLE 2.1
Major Producers of Reinforced Thermoplastic Materials in Europe (including UK)

Country	Company	Materials
France	Aquitaine Total Organico	Nylon 11, nylon 6
	Doneco (LNP Licensee)	All[a]
	Rhone Poulenc	Nylon 6.6, modified nylon 6.6, nylon 6.10
Germany	BASF	Nylon 6, nylon 6.6, acetal, styrene/acrylonitrile, polybutylene terephthalate
	Bayer	Nylon 6, polycarbonate, acrylonitrile/butadiene/styrene polymer
	DuPont	Nylon, acetal
	Hoechst	Acetal, polyethylene terephthalate, polybutylene terephthalate, polypropylene
	Chemische Werke Hüls	Polypropylene, nylon 12
Holland	AKZO	Polybutylene terephthalate, polyethylene terephthalate, nylon 6
	General Electric	Modified polyphenylene oxide, polycarbonate
	Shell	Polypropylene
Italy	Snia Viscosa	Nylon 6, nylon 6.6, nylon/polyester
	Lati	All[a]
	Montedison SpA	Nylon 6.6, nylon 6, styrene/acrylonitrile, polystyrene, acrylonitrile/butadiene/styrene polymer, polypropylene
Switzerland	Ciba Geigy (Fiberfil Licensee)	All[a]
	Emser Werke A.G.	Nylon 6, nylon 12
UK	ICI	Nylon 6.6, polybutylene terephthalate, polypropylene
	James Ferguson & Sons Ltd (LNP Licensee)	All[a]
	TBA Industrial Products Ltd	Polypropylene, polystyrene, nylon 6, polycarbonate
	Shell Chemical Co.	Polypropylene
	Polymon Developments Ltd	Polypropylene

[a] A wide range of reinforced polymers.

TABLE 2.2

Grades of Reinforced Thermoplastics Produced and Marketed in Europe (including UK)

Supplier	Trade name	Grade	Polymer	Reinforcement	Other additives and content
Aquitaine Total Organico (France)	Rilsan	ZM30	Nylon 11	30% glass fibre	—
		ZM23 G9	Nylon 11	23% glass fibre	Graphite
		ZM43 G9	Nylon 11	43% glass fibre	Graphite
		ZM30 Blk W3	Nylon 11	30% glass fibre	Flame retardant additives
		ZM30 Blk TL	Nylon 11	30% glass fibre	Heat and light stabilisers
		BMN G8	Nylon 11	—	Graphite
		BMN Y	Nylon 11	—	MoS_2
		BMN G9	Nylon 11	—	Higher loading graphite
		BMNY BZ TL	Nylon 11	Bronze beads	MoS_2, stabilised
		B-UM 30	Nylon 11	30% glass beads	—
	Orgamide	ZM 15 0	Nylon 6	15% glass fibre	—
		ZM 30 0	Nylon 6	30% glass fibre	—
		ZM 30 Blk 990	Nylon 6	30% glass fibre	—
		ZM 30 Blk T	Nylon 6	30% glass fibre	Heat stabilised
		ZM 23 G9	Nylon 6	23% glass fibre	Graphite
Rhone Poulenc (France)	Technyl	A216Y16	Nylon 6.6	33% glass fibre (short)	—
		A216Y15	Nylon 6.6	33% glass fibre (long)	—
		A216Y2	Nylon 6.6	—	Graphite
		A216Y10	Nylon 6.6	—	MoS_2

Table 2.2—*continued*

Supplier	Trade name	Grade	Polymer	Reinforcement	Other additives and content
		A216Y17	Nylon 6.6	33% glass fibre	Lubricant
		B216Y16	Modified nylon 6.6	33% glass fibre	—
		D316Y16	Nylon 6.10	33% glass fibre	—
BASF (Germany)	Luran	KR 2517	SAN	35% glass fibre	—
	Ultramid	B3WG5	Nylon 6	25% glass fibre	Stabilised
		B3WG6	Nylon 6	30% glass fibre	Stabilised
		B3WG7	Nylon 6	35% glass fibre	Stabilised
		B3WG10	Nylon 6	50% glass fibre	Stabilised
		KR 1346/203	Nylon 6	—	20% special chalk also stabilised
		KR 1346/303	Nylon 6	—	30% special chalk also stabilised
		A3 HG5	Nylon 6.6	25% glass fibre	Stabilised
		A3 XG5	Nylon 6.6	25% glass fibre	Flame retardant
		A3 WG5	Nylon 6.6	25% glass fibre	Stabilised
		A3 WG6	Nylon 6.6	30% glass fibre	Stabilised
		A3 WG7	Nylon 6.6	35% glass fibre	Flame retardant
		A3 WG10	Nylon 6.6	50% glass fibre	Stabilised
Bayer (Germany)	Durethan	BKV30	Nylon 6	30% glass fibre	—
		BKV30H	Nylon 6	30% glass fibre	Stabilised
		BKV30 ZH	Nylon 6	30% glass fibre	MoS_2, stabilised
		BK 31G	Nylon 6	—	Graphite
		BK 31Z	Nylon 6	—	MoS_2
	Makrolon	8020	Polycarbonate	20% glass fibre	—

Table 2.2—*continued*

Supplier	Trade name	Grade	Polymer	Reinforcement	Other additives and content
Bayer (Germany) —*contd.*		8030	Polycarbonate	30% glass fibre	—
		8320	Polycarbonate	20% glass fibre	Flame retardant
		8340	Polycarbonate	35% glass fibre	Flame retardant
		9310	Polycarbonate	10% glass fibre	Flame retardant
	Novadur	PH/GV	ABS	Glass fibre	—
Hoechst (Germany)	Hostaform C	CVP9023 GV	Acetal	25% glass fibre	—
		CVP9024M	Acetal	—	MoS_2
		CVP9024 TF	Acetal	—	PTFE
		CVP9024K	Acetal	—	Lubricant
	Hostalen PP	PPN VP 7780 GV	Polypropylene	20% glass fibre	—
		PPN VP 7190 TV	Polypropylene	Talc	—
		PPT VP 7090 AV	Polypropylene	Asbestos	—
		PPN VP 7180 TV	Polypropylene	—	Chalk 20%
		PPR VP 7352 FL	Polypropylene	—	Flame retardant
		PPN VP 7790 GVI	Polypropylene	30% glass fibre	—
		PPN VP GV2	Polypropylene	30% glass fibre	—
	Hostadur	KVP 8022 GV1	Polyester (PET)	20% glass fibre	—
		BVP 7600 GV2	Polyester (PBT)	30% glass fibre	—
Chemische Werke Hüls (Germany)	Vestamid	L1930	Nylon 12 (copolymer)	30% glass fibre	Black colour
		X2363	Nylon 12 (copolymer)	30% glass beads	—
	Vestolen P	P5232G	Polypropylene	20% glass beads	—
		P5232T	Polypropylene	Talc	—
		P5272T	Polypropylene	Talc	—

Table 2.2—*continued*

Supplier	Trade name	Grade (Injection)	Grade (Extrusion)	Polymer	Reinforcement	Other additives and content
AKZO (Holland)	Arnite	A300	X531	Polyester (PET)	18% glass fibre	—
		A301	X511	Polyester (PET)	18% glass fibre	8% graphite
		A350	X530	Polyester (PET)	18% glass fibre	TiO_2
		A340	X609	Polyester (PET)	36% glass fibre	—
		A330 SE-0	X632	Polyester (PET)	33% glass fibre	Flame retardant
		TV4 240 X		Polyester (PBT)	20% glass fibre	—
		TV4 260 SX		Polyester (PBT)	30% glass fibre	Flame retardant
		TV4 270 X		Polyester (PBT)	35% glass fibre	—
	Akulon	K2-62V		Nylon 6	30% glass fibre	—
		K2-62VX608		Nylon 6	15% glass fibre	—
		K2-62VX518		Nylon 6	20% glass fibre	—
		K2-62VX560		Nylon 6	25% glass fibre	—
		K2-62VX418		Nylon 6	35% glass fibre	—
Shell Chemicals (UK and Holland)	Carlona P	PLZ 532		Polypropylene	Talc	—
Snia Viscosa (Italy)	Sniavitrid	ASN 27/185		Nylon 6	18·5% glass fibre	—
		ASN 27/50		Nylon 6	25% glass fibre	—
		ASN 27/350		Nylon 6	35% glass fibre	—
		ASN 27/450		Nylon 6	45% glass fibre	—
		ASN 27/250 1		Nylon 6	25% glass fibre	Flame retardant
		ASN 27/250 AM		Nylon 6	25% asbestos	—
		SSD 185		Nylon 6.6	18·5% glass fibre	—

Table 2.2—*continued*

Supplier	Trade name	Grade	Polymer	Reinforcement	Other additives and content
Snia Viscosa (Italy)—*contd.*		SSD 250	Nylon 6.6	25% glass fibre	Also with flame retardant (sub-grades 250–1 and 350–1)
		SSD 350	Nylon 6.6	35% glass fibre	
		ASP 50/25	Nylon/polyester (copolymer)	25% glass fibre	
		SSD 300B	Nylon/polyester (copolymer)	30% glass beads	—
		SSD 400B	Nylon/polyester (copolymer)	40% glass beads	—
		ASN 27/300/B	Nylon 6	30% glass beads	—
		ASN 27/400/B	Nylon 6	40% glass beads	—
Lati SpA (Italy)	Lastirol LV	Various	Polystyrene	25–40% glass fibre	Flame retardant grades are available
	Lastil LV	Various	SAN	25–40% glass fibre	—
	Lastilac LV	Various	ABS	5–22% glass fibre	Flame retardant grades are available
	Latene 100 LV	Various	Polyethylene HD	20–30% glass fibre	Flame retardant grades are available
	Latene 120 LV	Various	Polypropylene	20–75% glass fibre	Flame retardant grades are available
	Latamid 6 LV	Various	Nylon 6	30–50% glass fibre	Lubricated grades are available
	Latamid 66 LV	Various	Nylon 6.6	30–50% glass fibre	Flame retardant and lubricated grades are available

Table 2.2—*continued*

Supplier	Trade name	Grade	Polymer	Reinforcement	Other additives and content
	Latamid 12 LV	Various	Nylon 12	20–30% glass fibre	Lubricated grades are available
	Laster B LV	Various	Polyester (PET)	20–30% glass fibre	Flame retardant and lubricated grades are available
	Latan LV	Various	Acetal	15–30% glass fibre	Lubricated grades are available
	Latilon LV	Various	Polycarbonate	20–30% glass fibre	Flame retardant grades are available
	Laril LV	Various	Modified PPO	20–30% glass fibre	Flame retardant grades are available
	Lasulf LV	Various	Polysulphone	20–30% glass fibre	—
Montedison SpA (Italy)	Renyl MVR3C	Various	Nylon 6	30% glass fibre	—
	Moplen		Polypropylene	Asbestos, talc, 30% glass fibre	—
Emser Werke AG (Switzerland)	Grilon	PV-3H	Nylon 6	30% glass fibre	Heat stabilised
		PK 5H	Nylon 6	Glass beads	Heat stabilised
	Grilamid	LV 3H	Nylon 12	30% glass fibre	—
		LK 5H	Nylon 12	Glass beads	—

Table 2.2—*continued*

Supplier	Trade name	Grade	Polymer	Reinforcement	Other additives and content
ICI Ltd (UK)	Maranyl	A190	Nylon 6.6	33% glass fibre	—
		A192	Nylon 6.6	33% glass fibre	Easy flow A190
		A108	Nylon 6.6	—	MoS_2/graphite
		A198	Nylon 6.6	33% glass fibre	Graphite
		AD180	Nylon 6.6	33% glass fibre	—
		AD182	Nylon 6.6	33% glass fibre	Easy flow AD180
		AD197	Nylon 6.6	33% glass fibre	Flame retardant
		AD390	Nylon 6.6	40% glass fibre	—
		AD590	Nylon 6.6	33% glass fibre	Extrusion grade
		AD790	Nylon 6.6	Glass fibre/beads	—
		F193	Nylon 6	Undisclosed	—
	Propathene	HW70GR	Polypropylene	25% glass fibre	—
		22/44 Blk 9040	Polypropylene	Asbestos	—
		PXC4 02 and 6602	Polypropylene	40% talc	—
	Deroton	TGA50	Polyester (PBT)	30% glass fibre	—
		TGR80	Polyester (PBT)	30% glass fibre	Flame retardant
Turner Brothers Asbestos Co. Ltd (UK)	Arpylene APP	N 2025	Polypropylene	25% asbestos	—
		N 1040	Polypropylene	40% asbestos	—
		N 2045	Polypropylene	45% asbestos	—
		N 2240	Polypropylene	40% asbestos	Flame retardant
	Arpylene TPP	N 2025	Polypropylene	25% talc	—
		N 2040	Polypropylene	40% talc	—
		N 3080 (Master-batch)	Polypropylene	80% talc	—
	Arpylene APS	N 1025	Polystyrene	25% asbestos	—

Table 2.2—*continued*

Supplier	Trade name	Grade	Polymer	Reinforcement	Other additives and content
		N 1325	Polystyrene	25% asbestos	Flame retardant
	Arpylene GPC	N 1030	Polycarbonate	30% glass fibre	—
	Arpylene GPA	N 1030	Nylon 6	30% glass fibre	—
Polymon Developments Ltd (UK)	Formid	HMT	Polypropylene	Talc	—
		HFA	Polypropylene	Asbestos	—
		FRP	Polypropylene	—	Flame retardant
		FRA	Polypropylene	Asbestos	Flame retardant

The Japanese reinforced polyester (FR-PET of the Teijin Co.) is marketed in the UK by the agents Kingsley & Keith.

their materials (which they resell, manufacture locally or produce under licence) are listed in the section on the USA (Section 2.2).

2.2 THE USA AND JAPAN

The arrangement of the information in this section follows the pattern of Section 2.1, with two exceptions. Firstly, as has been mentioned, the details of the American reinforced thermoplastic materials (Table 2.4) include also those materials which are sold in Europe by subsidiaries or licensees. Secondly, for Japan, only a general list of manufacturers and products is given without trade names and grade details. The list (Table 2.5) differs from the pattern of Tables 2.1 and 2.3 in that all major Japanese polymer producers have been included with an indication of which ones market reinforced grades.

TABLE 2.3
Major Producers of Reinforced Thermoplastics Materials in the USA

Company	Materials
Fiberfil Div. Dart Industries Inc.	All: a full range
Liquid Nitrogen Processing Corp.	All: incl. polytetrafluoroethylene, and ethylene/tetrafluoroethylene and fluorinated ethylene/propylene polymers
Thermofil	Most
Allied Chemical Co.	Nylon 6
Celanese Corp.	Acetal, polybutylene terephthalate, nylon 6
DuPont	Nylon 6.6, acetal, ethylene/tetrafluoroethylene polymer, nylon 6.12
General Electric Plastics Div.	Polycarbonate, polybutylene terephthalate, modified polyphenylene oxide
Eastman Kodak	Polybutylene terephthalate
Nypel Inc.	Nylon 6, nylon 6.6
Wellman Inc.	Nylon 6, nylon 6.6
Polymer Corp.	Nylon 6.6
Exxon Chemical Co.	Polypropylene
Amoco	Polypropylene
Phillips Petroleum	Polyphenylene sulphide

Until recently Ethyl Corporation (Polymer Division) manufactured a PVC compound reinforced with 20% glass fibre (Ethyl 8230). Currently they supply the PVC polymer (Ethyl 7042 Natural) for mixing with glass fibre at the injection moulding machine.

Hercules Inc. similarly supply a grade of polypropylene (Profax PC-072) for mixing by the customer with glass fibre.

TABLE 2.4

Grades of Reinforced Thermoplastics Produced and Marketed in the USA

Supplier	Trade name	Grade	Polymer	Reinforcement	Other additives and content
Fiberfil Inc., a division of Dart Industries	Nylafil[a]	G-1/30, G-10/40	Nylon 6.6	30 and 40% glass fibre	—
	Nylafil	G-2/30, G-12/40	Nylon 6	30 and 40% glass fibre	—
	Nylafil	G-3/30, G-13/40	Nylon 6.10	30 and 40% glass fibre	—
	High lubricity Nylafil	G-1/30MS/5,	Nylon 6.6	30% glass fibre	MoS_2
		G-1/30TF/22	Nylon 6.6	30% glass fibre	PTFE
	Nylasar[a]	J-1/30, J-10/40	Nylon 6.6	30 and 40% glass fibre	—
	Nylasar	J-2/30. J-12/40	Nylon 6.10	30 and 40% glass fibre	—
	Nylasar	J-3/30, J-13/40	Nylon 6	30 and 40% glass fibre	—
	High lubricity Nylasar	J-1/30MS/5,	Nylon 6.6	30% glass fibre	MoS_2
		J-1/30TF/22	Nylon 6.6	30% glass fibre	PTFE
	Nylaglas[a]	S-1/30, S-2/3, S-3/30	Nylons 6.6, 6.10 and 6 respectively	30% glass fibre	—
	Nylode Plaslode	NY-1/ASF/20	Nylon 6.6	20% asbestos	—
	Styrafil	G-37/20, G-30/30, G35/35	Polystyrene	20, 30 and 35% glass fibre	—
	Styrasar	J-30/20, J-30/30	Polystyrene	20 and 30% glass fibre	
	Styraglas	S-30/20	Polystyrene	20% glass fibre	—
	Acrylafil	G-47/20, G-40/35	SAN	20 and 35% glass fibre	—
	Acrylasar	J-40/20	SAN	20% glass fibre	—
	Acrylaglas	S-40/20, S-40/35	SAN	20 and 35% glass fibre	—
	Absafil	G-1200/20, G-1200/40	ABS	20 and 40% glass fibre	—
	Absasar	J-1200/20	ABS	20% glass fibre	—
	Absaglas	S-1200/20	ABS	20% glass fibre	—

Table 2.4—*continued*

Supplier	Trade name	Grade	Polymer	Reinforcement	Other additives and content
Fiberfil Inc. —*contd.*	Polycarbafil	G-50/20, G-50/40	Polycarbonate	20 and 40% glass fibre	—
	High lubricity Polycarbafil	G-50/20TF/22, G-50/40TF/12	Polycarbonate	20 and 40% glass fibre	PTFE
	Carbasar	J-50/20, J-50/30, J-50/40	Polycarbonate	20, 30 and 40% glass fibre	—
	Carbaglas	S-50/20	Polycarbonate	20% glass fibre	—
	Sulfil	G-1500/10, G-1500/20	Polysulphone	10 and 20% glass fibre	—
	Sulfasar	J-1500/20	Polysulphone	20% glass fibre	—
	Profil	G-60/20, G-60/40	Polypropylene	20 and 40% glass fibre	—
	Propylsar	J-60/20, J-60/30,	Polypropylene	20 and 30% glass fibre	—
		J-60/20/HS	Polypropylene	20% glass fibre	Heat stabiliser
	Ethofil	G-90/20, G-90/40	HD Polyethylene	20 and 40% glass fibre	—
	Ethosar	J-90/20, J-90/30	HD Polyethylene	20 and 40% glass fibre	—
	Formaldafil	G-80/20	Acetal (copolymer)	20% glass fibre	—
	Formaldafil	G-81/40	Acetal (homopolymer)	40% glass fibre	—
	Formaldasar	J-80/20, J-80/40	Acetal (copolymer)	20 and 40% glass fibre	—
	Urafil	G-100/20	Polyurethane	20% glass fibre	—
	Urasar	J-100/20	Polyurethane	20% glass fibre	—
Liquid Nitrogen Processing Corp.	'Thermocomp'	AF series	ABS	20–40% glass fibre	—
		BF series	SAN	20–40% glass fibre	—
		CF series	Polystyrene	20–40% glass fibre	—
		DF series	Polycarbonate	20–40% glass fibre	PTFE lubricated grades 5–20% available

[a] Suffixes: FIL = long fibres (coaxial); SAR = medium length fibres; GLAS = short, random fibres.

Table 2.4—*continued*

Supplier	Trade name	Grade	Polymer	Reinforcement	Other additives and content
	'Thermocomp'	EF series	Polybutene	30% glass fibre	—
		FF series	Polyethylene	20–40% glass fibre	—
		GF series	Polysulphone	20–40% glass fibre	PTFE lubricated grades 5–20%
		HF series	Nylon 11	30% glass fibre	—
		IF series	Nylon 6.12	23–35% glass fibre	—
		KF series	Acetal	20–40% glass fibre	PTFE lubricated grades 5–20%
		JF series	Polyether sulphone	20–40% glass fibre	—
		MF series	Polypropylene	20–40% glass fibre	—
		OF series	Polyphenylene sulphide	20–40% glass fibre	PTFE lubricated grades 5–20%
		PF series	Nylon 6	10–60% glass fibre	PTFE lubricated grades 5–20%
		QF series	Nylon 6.10	10–60% glass fibre	PTFE lubricated grades 5–20%
		RF series	Nylon 6.6	10–60% glass fibre	PTFE lubricated grades 5–20%
		SF series	Nylon 12	30% glass fibre	—
		TF series	Polyurethane	20–40% glass fibre	—
		UF series	ETFE	20–40% glass fibre	—
		VF series	Polyvinyl chloride	15% glass fibre	—
		WF series	Polyester	20–40% glass fibre	PTFE lubricated grades 5–20%
		YF series	Polyimidal	20–40% glass fibre	—
		ZF series	Modified PPO	20–40% glass fibre	—

Table 2.4—*continued*

Supplier	Trade name	Grade	Polymer	Reinforcement	Other additives and content
Liquid Nitrogen Processing Corp.—*contd.*	'Thermocomp'	GC series	Polysulphone	20–40% carbon fibre	—
		OC series	Polyphenylene sulphide	20–40% carbon fibre	—
		RC series	Nylon 6.6	20–40% carbon fibre	—
		UC series	ETFE	20–40% carbon fibre	—
		WC series	Polyester	20–40% carbon fibre	—
		RM series	Nylon 6.6	Mineral-filled	—
		RF B series	Nylon 6.6	30% glass beads	—
		PF B series	Nylon 6	30% glass beads	—
		QF B series	Nylon 6.10	30% glass beads	—
		MF B series	Polypropylene	30% glass beads	—
		DF B series	Polycarbonate	30% glass beads	—
		FF B series	HD polyethylene	30% glass beads	—
		AW[b] series	ABS	30% Fybex (potassium titanate whiskers)	
		MW[b] series	Polypropylene	25% Fybex (potassium titanate whiskers)	—
		RW[b] series	Nylon 6.6	25% Fybex (potassium titanate whiskers)	—
		GW[b] series	Polysulphone	25% Fybex (potassium titanate whiskers)	—
Allied Chemical	Plaskon	8230	Nylon 6	6% glass fibre	—
		8231	Nylon 6	14% glass fibre	—
		8233	Nylon 6	30% glass fibre	—

[b] Not available since 1974, as the production of Fybex has been discontinued by DuPont.

Table 2.4—*continued*

Supplier	Trade name	Grade	Polymer	Reinforcement	Other additives and content
Celanese Corp.	Celanex	3300	Polyester (PBT)	30% glass fibre	—
		3310	Polyester (PBT)	30% glass fibre	FR (Highest heat resistance)
		3210	Polyester (PBT)	30% glass fibre	FR (Highest heat resistance)
	Celcon (sold as Kematal in UK)[c]	M90 GC25	Acetal copolymer	25% glass fibre	—
		M90 105	Acetal copolymer	—	22% PTFE
DuPont	Delrin	570X NC-000	Acetal homopolymer	20% glass fibre	—
		AF	Acetal homopolymer	—	PTFE fibres
	Zytel[e]	7010-13	Nylon 6.6	13% glass fibre	—
		7010-13C[d]	Nylon 6.6	13% glass fibre	—
		7010-33	Nylon 6.6	33% glass fibre	—
		7030-33	Nylon 6.6	33% glass fibre	Stabilised
	Zytel	7040-33	Nylon 6.6	33% glass fibre	Hydrolysis-resistant
		7110-13	Mod. high impact nylon 6.6	13% glass fibre	—
		7110-13C[d]	Nylon 6.6	13% glass fibre	—
		7110-33	Nylon 6.6	33% glass fibre	—
		7910-13	Nylon 6.6 (modified)	13% glass fibre	—
		7910-33	Nylon 6.6 (modified)	33% glass fibre	—
	Tefzel[e]	70G-25	ETFE	25% glass fibre	—

[c] Kematal[R] is a registered trade mark of the Celanese Corporation (USA) the parent company of Amcel Ltd in the UK.
[d] Cube blend.
[e] Zytel[R] and Tefzel[R] are registered trade marks of the DuPont Company.

Table 2.4—*continued*

Supplier	Trade name	Grade	Polymer	Reinforcement	Other additives and content
General Electric Plastics Div.	Lexan	3412	Polycarbonate	20% glass fibre	—
		3414	Polycarbonate	40% glass fibre	—
		500	Polycarbonate	10% glass fibre	Flame retardant
	Noryl	GFN2	Toughened polystyrene/modified PPO	20% glass fibre	—
		GFN3	Toughened polystyrene/modified PPO	30% glass fibre	—
		GFN2 SE-1	Toughened polystyrene/modified PPO	20% glass fibre	Flame retardant
		GFN3 SE-1	Toughened polystyrene/modified PPO	30% glass fibre	Flame retardant
	Valox	420	Polyester (PBT)	25% glass fibre	—
		420-SE0	Polyester (PBT)	25% glass fibre	Flame retardant
Nypel Inc.	Nyreg	6610	Nylon 6.6	10% glass fibre	—
		6620	Nylon 6.6	20% glass fibre	—
		6630	Nylon 6.6	30% glass fibre	—
		6640	Nylon 6.6	40% glass fibre	—
		6010	Nylon 6	10% glass fibre	—
		6014	Nylon 6	14% glass fibre	—
		6020	Nylon 6	20% glass fibre	—
		6030	Nylon 6	30% glass fibre	—
		6040	Nylon 6	40% glass fibre	—

Table 2.4—*continued*

Supplier	Trade name	Grade	Polymer	Reinforcement	Other additives amd content
Exxon Chemical Co.		CD 207C	Polypropylene	20% talc	—
		CD 119D	Polypropylene	40% talc	—
		CD 116D	Polypropylene	40% asbestos	—
Thermofil Inc.		F.G.	Nylon 6 and 6.6	5–40% glass fibre	Most grades supplied with additives, including PTFE and MoS_2, flame retardants and conductive fillers
		B.G.	Nylon 6 and 6.6	15–40% glass beads	
		F.G.	Nylon 6.10	10–40% glass fibre	
		F.G.	Nylon 12	10–50% glass fibre	
		F.G.	Polypropylene	10–40% glass fibre	
		F.G.	Polypropylene	10–40% talc	
		F.G.	HD polyethylene	10–30% glass fibre	
		F.G.	ABS	10–30% glass fibre	
		F.G.	SAN	10–35% glass fibre	
		F.G.	Polystyrene	10–30% glass fibre	
		F.G.	Acetal	10–30% glass fibre	
		F.G.	Polycarbonate	10–40% glass fibre	
		F.G.	Polysulphone	10–30% glass fibre	
		F.G.	Modified PPO	10–40% glass fibre	
		F.G.	Polyester	10–30% glass fibre	

TABLE 2.5
Major Producers of Thermoplastics in Japan
(Polymers marked with an asterisk are offered in reinforced as well as unreinforced grades)

Name of company	Polymer
Asahi Glass Co., Ltd.	PVC*, TFE, ETFE
Mitsubishi Chemical Industries Ltd.	N6*, HDPE
Mitsubishi Petrochemical Co., Ltd.	PP*, HDPE, PS
Mitsubishi Monsanto Chemical Co., Ltd.	PS*, SAN*, ABS*, PVC
Mitsubishi Gas-Chemical Co., Inc.	PC*
Mitsubishi Rayon Co., Ltd.	PMMA, ABS, PP*, PBT*
Mitsui Toatsu Chemicals, Inc.	SAN, ABS, PS, HDPE, PP*, PVC
Sumitomo Chemical Co., Ltd.	ABS, HDPE, PP, PVC, PMMA
Asahi Chemical Industry Co., Ltd.	HDPE, N6.6*, PA*, PMMA
Asahi Dow Chemical Co., Ltd.	PS*, SAN*, ABS
Toray Industries Co., Ltd.	SAN*, ABS*, PP*, N6*, N6.6*, N6.10, N12, PBT*
Teijin Chemical Co., Ltd.	PC*, PBT*, PET*
Ube Industries, Ltd.	N6*, N6.6, PP, ABS
Toyobo Co., Ltd.	N6
Unitica Ltd.	N6
Kanegafuchi Chemical Industry Co., Ltd.	PS, SAN, ABS, PVC
Nippon Steel Chemical Co., Ltd.	PS, SAN*
Chisso Corporation	HDPE, PP, PVC
Daicel Ltd.	SAN*, ABS*
Polyplastics Co., Ltd.	PA*
Idemitsu Petrochemical Co., Ltd.	PS, ABS, SAN, HDPE
Nippon Zeon Co., Ltd.	ABS, PVC
The Electro-Chemical Industrial Co., Ltd.	PS, ABS, PVC
Sekisui Chemical Co., Ltd.	PS, PVC
Dainippon Ink and Chemicals Inc.	PS
Japan Synthetic Rubber Co., Ltd.	ABS
The Nisseki Plastic Chemical Co., Ltd.	HDPE
Chubu Chemical Co., Ltd.	HDPE
Nissan Chemical Industries, Ltd.	HDPE, PVC
Tokuyama Soda Co., Ltd.	PP
Showa Denko K. K.	PP, HDPE
Kureha Chemical Industry Co., Ltd.	PVC, PVDF
Nippon Carbide Industries Co., Ltd.	PVC
Nisshin Chemical Industry Co., Ltd.	PVC
Gumma Chemical Industry Co., Ltd.	PVC
Shinetsu Chemical Industry Co., Ltd.	PVC
Tekkosha Co., Ltd.	PVC
Toagosei Chemical Industry Co., Ltd.	PVC
Daikin Kogyo Co., Ltd.	PTFE
Mitsui Fluorochemical Co., Ltd.	PTFE, FEP
Tohkai Kogyo Rubber Co., Ltd.	PP, ABS, N6, PS

CHAPTER 3

POLYMERS USED IN THE PREPARATION OF REINFORCED THERMOPLASTIC MATERIALS

3.1 BASE POLYMERS: SELECTION AND GENERAL POINTS

The polymeric components of reinforced thermoplastics are commonly referred to as 'base polymers'. They are normally moulding grades of commercial polymers. Where several moulding grades of a particular polymer are available, and in the absence of special requirements, the selection will be made in the light of two main considerations which are essentially conflicting: the highest molecular weight grade within the range is desirable for the best physical properties of the polymer (and hence the polymer matrix of composite to be produced) whilst ease of processing increases with decreasing molecular weight. In practice the choice is a compromise frequently considerably biased towards the higher molecular weight for good physical properties, because processing can be assisted by incorporation of lubricants (e.g. aluminium stearate in nylon or various lubricants in rigid PVC) or by addition of an easily blending polymer (e.g. the lower melting nylon 6 to aid in the processing of nylon 6.6).

Some 30 polymers have been used to date as base polymers in commercial reinforced thermoplastic materials. These polymers are listed in the tables of Section 3.2 with general data on their characteristics and sources.

Some physical and processing properties including those which are the most strongly affected by the introduction of reinforcing filler into the base polymer are shown in the tables of Section 3.3. The effects of reinforcement, with 30% glass fibre by weight, are illustrated by a comparison of typical numerical values for the unfilled and filled polymer.

3.2 THE BASE POLYMERS OF REINFORCED THERMOPLASTIC MATERIALS

The data on the polymers are grouped according to the main chemical types in Tables 3.2(a)–3.2(i). Under the heading 'Appearance' in the tables (column 4) the descriptions given relate to the 'natural' polymer, i.e. without pigments, colourants, etc. The softening point data where given in column 5 of the tables are the Vicat softening points, i.e. the temperatures at which specified needle penetration occurs when specimens are subjected to specified test conditions according to the Vicat method (*cf.* e.g. ASTM Standard D 1525).

3.2.1 Polyamides

$$\left[\text{HN}(\text{CH}_2)_n\text{NH—OC}(\text{CH}_2)_m\text{CO} \right] \tag{I}$$

or

$$\left[\text{HN}(\text{CH}_2)_{n_1}\text{CO} \right] \tag{II}$$

The polyamides (nylons) are among the best engineering thermoplastics. They are tough, resistant to abrasion and strong in comparison with many thermoplastics. Their resistance to chemicals and microbiological attack is very good.

The main performance differences between the different chemical types are in the melting point, softening temperature and moisture absorption. There are also differences in strength and modulus in the fully water-saturated condition because water has a plasticising action on the polyamide. Water absorption at room temperature and some of its effects on dimensions and stiffness are shown in Table 3.1.

Reinforcement can increase three-fold the temperature for heat distortion under load of these materials, and similarly improve their compressive strength, tensile strength and stiffness. World-wide, glass reinforced nylons are probably the most improved single group of reinforced thermoplastics. Compared with other engineering plastics the volume of production of nylon 6.6 and nylon 6 is very high. This is because these polymers are produced in high quantities for textile use. This situation is favourably reflected in the price of these two nylons which are comparatively cheap relative to both the other polyamides and other engineering polymers of comparable property levels.

3.2.2 Acetal Polymers and Copolymers

$$\left[\text{CH}_2\text{O} \right]$$

TABLE 3.1
Effect of Water on Nylons
(Based in part on data from *Modern Plastics Encyclopedia*, issue 1971–72, reprinted by permission of Modern Plastics Encyclopedia, McGraw-Hill Inc.)

	Water absorption (%)		Dimensional change (%)		Flexural modulus (psi)		
Type	50% RH	Saturation	50% RH	Saturation	Dry	50% RH	Saturation
6.6	2·5	9·0	0·6	2·8	410 000	175 000	85 000
6	2·7	9·5	0·7	3·0	395 000	140 000	70 000
6.10	1·5	3·5	0·2	0·7	280 000	160 000	100 000
6.12	1·3	3·0	0·2	0·6	290 000	180 000	120 000
11	0·8	1·9	0·1	0·4	192 000	142 000	138 000
12	0·7	1·4	0·1	0·3	204 000	175 000	148 000

The nylons listed in Table 3.1 are all aliphatic, crystalline polymers (*cf.* Table 3.2(a)). A non-crystalline moulding polyamide (Trogamid T), based on trimethylhexamethylene diamine and terephthalic acid, is also available (from Dynamit Nobel). It is not used commercially in reinforced form. Its amorphous nature makes Trogamid T the only clear, transparent commercial polyamide. This property is impaired by incorporation of fillers.

High strength, stiffness, resistance to creep, resilience and dimensional stability over a considerable temperature range, combined with high fatigue endurance, are the outstanding properties of this group of polymers, making them particularly useful in engineering applications. Their chemical resistance and resistance to water is also very good. The further improvement of strength and stiffness with fibrous reinforcement (*cf.* Section 3.3) makes the polyacetals competitive with metals for certain applications (*cf.* Chapter 6).

The main disadvantageous characteristics of the acetal polymers are a tendency to decomposition at high temperatures, particularly in the homopolymer, as, e.g. in processing, requiring stabilisation, flammability, and not particularly ready adhesion to glass fibre reinforcement.

Acetal homopolymer can be distinguished from the copolymer by heating in triethanolamine at 200°C. The homopolymer is much less resistant to decomposition in these conditions: it turns brown and then almost black with evolution of formaldehyde after a few minutes. The copolymer may develop yellowish colour, but is otherwise unchanged. The test is simple, but normal precautions should be observed in view of the flash point (185°C) and strong alkalinity of triethanolamine, and toxicity of formaldehyde vapour.[1]

TABLE 3.2(a)
Polyamides—General Data

Polymer	Chemical structure details	Fine structure	Appearance	Melting point (Tc)[a] and/or softening point (Ts)	Examples of commercial names and grades
Nylon 6.6 (polyhexamethylene adipamide)	Essentially linear chains Type I: $n = 6$, $m = 4$	Crystalline, can be oriented	Practically colourless to slight yellow. Translucent (degree varies with degree of crystallinity)	$Tc = 264°C$ $Ts = 258°C$	Maranyl (ICI) Zytel 101 (DuPont) Sniamid SSD (Snia Viscosa)
Nylon 6 (polycaprolactam)	Essentially linear chains Type II: $n_1 = 5$	Crystalline, can be oriented	As above	$Ts = 160–220°C$ $Tc = 225°C$	Ultramid B-3 (BASF) Grilon (Emser Werke) Akulon (AKU) Zytel (DuPont)
Nylon 6.10 (polyhexamethylene sebacamide)	Essentially linear chains Type I: $n = 6$, $m = 8$	Crystalline, can be oriented	As above	$Tc = 222°C$	Zytel 31 (DuPont) Maranyl B100 (ICI)
Nylon 6.12 (polyhexamethylene dodecanoamide)	Essentially linear chains Type I: $n = 6$, $m = 10$	Crystalline, can be oriented	As above	Melting point (Fisher–Johns) = 208–215°C	Zytel 151 (DuPont)
Nylon 11 (polyundecano-amide)	Essentially linear chains Type II: $n_1 = 10$	Crystalline, can be oriented	As above	$Tc = 185°C$	Rilsan (Aquitaine-Organico)
Nylon 12 (polylaurolactam) N.B.: Lowest water absorption and density among nylons	Essentially linear chains Type II: $n_1 = 11$	Crystalline, can be oriented	As above	$Tc = 178–180°C$ $Ts = 168–175°C$	Grilamid (Emser Werke) Vestamid (Chemische Werke Hüls)

[a] The melting points are for the completely dry polymers.

TABLE 3.2(b)
Acetal Polymers and Copolymers—General Data

Polymer	Chemical structure details	Fine structure	Appearance	Melting point (Tc)[a] and/or softening point (Ts)	Examples of commercial names and grades
Polyacetal (polyformaldehyde or copolymer)	Homopolymer: virtually unbranched polyoxymethylene chains; Copolymer: linear chains as in homopolymer but containing sequences of adjacent C atoms	Highly crystalline, can be oriented	Translucent to opaque white	Homopolymer: $Tc = 180°C$ $Ts = 173°C$ Copolymer (Celcon): $Tc = 170°C$ $Ts = 168°C$	Homopolymer: Delrin (DuPont) Copolymer:[b] Celcon M90 (Celanese Plastics Co.) Kematal (Amcel)

[a] The melting points are for the completely dry polymers.
[b] Celcon and Kematal are in fact the same copolymer but different trade names are used in the UK and USA for commercial reasons.

3.2.3 Polyolefins

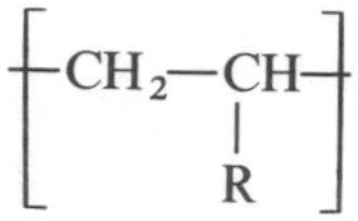

$$\left[\!-CH_2-\underset{\displaystyle R}{\underset{|}{CH}}-\right]$$

These materials, with the exception of poly-4-methylpentene-1, are essentially commodity plastics with comparatively poor 'engineering' properties. However the group represents the largest production and consumption volume in thermoplastics. The outstanding features of the run-of-the-mill polyolefins, i.e. polyethylene (high, medium and low density) and polypropylene, are good processability, chemical inertness, water resistance and good electrical properties combined with low density. It may be noted that low density polyethylene is not of commercial importance as a base polymer for reinforced products because its 'engineering' properties are poor and are not upgraded sufficiently by reinforcement. The stiffness, strength and distortion temperature of HD polyethylene and polypropylene are greatly improved by reinforcement, especially with glass fibre, and reach values comparable with those for the best (unreinforced) engineering plastics, e.g. nylon, polyester. The creep of polypropylene containing 30% glass fibre is very low.

The remaining two members of the group may be regarded as special cases. Thus poly-4-methylpentene-1, whose outstanding properties are high softening temperature and clarity, is a speciality material. Whilst reinforcement upgrades the mechanical properties, it obviously impairs the clarity of the polymer. Commercial reinforced grades are available but are not widely used. Originally ICI were virtually the sole supplier of this material under the trade name TPX. In 1973 the whole operation, including know-how, was transferred to Mitsui in Japan.

Like poly-4-methylpentene-1, polybutene is available commercially in reinforced grades but is not widely used. The base polymer has better creep characteristics than the main polyolefins.

Mention should also be made of the so-called ionomers, commercially represented by Surlyn (DuPont). These are ethylene copolymers with ionic inter-chain links. Glass-fibre-reinforced Surlyn sheets are manufactured, but they are of limited commercial importance and no data are therefore included in the tables of this section or Section 3.3.

3.2.4 Thermoplastic Polyesters

$$\left[-OC-C_6H_4-COO(CH_2)_2O-\right] \qquad (I)$$

TABLE 3.2(c)
Polyolefins—General Data

Polymer	Chemical structure details	Fine structure	Appearance	Melting point (Tc) and/or softening points (Ts)	Examples of commercial names and grades
Polyethylene (only high density polymer used in reinforced thermoplastics)	Partly branched chains, degree of branching decreasing with increasing density $R = H$	Crystalline, can be oriented	Translucent to opaque depending on amount and form of crystallinity	Low density: $Tc = 110°C$ $Ts = 85–87°C$ High density: $Tc = 135°C$ $Ts = 120–130°C$	Hostalen (Hoechst) Alathon (DuPont)
Polypropylene	Substantially linear chains: commercial polymers predominantly isotactic, small syndiotactic and atactic content $R = CH_3$	Crystalline, can be oriented	Transparent to translucent depending on degree and nature of crystallinity	$Tc = 165°C$ $Ts = 150°C$	Polypropylene (Amoco Chemicals Corp.) Propathene (ICI) Vestolen (Chemische Werke Hüls) Pro-fax PC-072 (Hercules Inc.)
Polybutene-1	Essentially linear chains, largely isotactic $R = C_2H_5$	Highly crystalline, can be oriented	Translucent to opaque depending on degree of crystallinity	$Tc = 140°C$	P6-003 (Mobil Chemical Co.)
Poly-4-methyl-pentene-1	Substantially linear chains $R = (CH_3)_2CHCH_2$	Crystalline	Transparent (crystalline and amorphous forms have the same refractive index)	$Tc = 245°C$ $Ts = 175°C$	TPX (originally ICI now Mitsui)

or

$$\left[\!-\mathrm{OC}\!\left\langle\bigcirc\right\rangle\!\mathrm{COO(CH_2)_4O}-\!\right] \qquad \text{(II)}$$

These are comparatively new moulding materials. Structurally they may be regarded as linear condensation products of terephthalic acid and linear glycols. Historically polyethylene terephthalate was the first thermoplastic polyester to be used as injection moulding material (Arnite, marketed by AKZO Plastics, Holland). Shortly afterwards the Japanese FR-PET appeared on the market but this is a glass-fibre-reinforced grade of polyethylene terephthalate and no corresponding unreinforced grade is available. These materials were introduced commercially in the mid-1960s. Towards the end of that decade the second generation of polyesters appeared, *viz.* polytetramethylene terephthalate, also referred to as polybutylene terephthalate. These materials provide a particularly good illustration of the elevation, through reinforcement, of the properties of a comparatively mediocre material to a very good and versatile engineering plastic. This is demonstrated by the difference of some 140°C between the temperature for deflection under load of unreinforced polybutylene terephthalate (about 50°C) and the same material reinforced with glass fibres (about 190°C). The tensile strength can also be doubled (from about 8000 to 16 000 psi).

With the exception of the Japanese FR-PET material and Arnite T, which are available only in reinforced form, all the other main thermoplastic polyesters are available in unreinforced as well as reinforced grades. An unfilled grade of Arnite T is expected to be produced shortly by AKZO Plastics bv.

The base polymer grades are:

Polyethylene terephthalate	*Polybutylene terephthalate*
Hostadur KVP 4022	Tenite 6 PRO
Arnite (several grades)	Deroton TAP 10
	Hostadur BVP 860
	Celanex 2001
	Valox 310

Most of these polymers are available at several levels of glass reinforcement and in glass-reinforced flame-retardant form.

With the exception of the low deflection temperature under load in the case of polybutylene terephthalate, the thermoplastic polyesters are in general good engineering plastics and become excellent when glass-reinforced. Whilst their tensile strength is somewhat lower than that of nylon 6.6, they possess good fatigue endurance and toughness. Their dimensional stability in the presence of moisture is much superior to that of the nylons because of the polyesters' considerably lower water absorption. The processability is very good enabling fast cycle times in

TABLE 3.2(d)
Thermoplastic Polyesters—General Data

Polymer	Chemical structure details	Fine structure	Appearance	Melting point (Tc)[a] and/or softening point (Ts)	Examples of commercial names and grades
Modified polyethylene terephthalate	Type I: Chains substantially linear	Crystalline, can be oriented. Amorphous, transparent grades also available (e.g. Hostadur AVP 4000)	Opaque white	$Tc = 255–258°C$	Hostadur KVP 4022 (Farbwerke Hoechst AG) Arnite A (AKZO) FR-PET (Teijin Ltd.)
Polytetra-methylene terephthalate	Type II: Chains substantially linear	Highly crystalline can be oriented	Translucent to opaque white	$Tc =$ approx. 224–225°C	Tenite 6PRO (Eastman Chemical) Deroton TAP 10 (ICI) Celanex 2001 (Celanese Plastics) Valox 310 (General Electric Co.) Hostadur BVP 860 (Farbwerke Hoechst AG) Arnite T (AKZO)

[a] The melting points are for the completely dry polymers.

moulding. Polybutylene terephthalate is considered to be particularly good in this respect.

Among the disadvantageous characteristics are a degree of notch sensitivity under impact and swelling by chlorinated hydrocarbon solvents, although the general chemical resistance is good. The normal grades are flammable but grades with the lowest flammability ratings (V-O) under the American Underwriters' Laboratories system are also available (*cf.* Chapter 9).

3.2.5 Polyarylates

The classification into sections in this chapter is one of convenience and is not solely and strictly based on chemical structure. It is convenient to group together the three 'polyarylates', two of which (polysulphone and polycarbonate) are based on bisphenol A, and the third on phenylene oxide. All are linear polymers and all are characterised by softening points which are high for thermoplastics. Bisphenol A polycarbonate is the oldest and longest-established of the three materials; it is also the one which gives water-white clarity, although it must be remembered that bisphenol A polycarbonate is crystallisable or can crystallise under the influence of heat and/or solvents, in which case opacification can develop. Polysulphone, whilst clear, has a yellow tinge. Polyphenylene oxide is commercially available as a blend with toughened polystyrene (Noryl) to facilitate processing. In addition to their high softening points, all three materials are strong and polycarbonate is particularly tough, especially in thin sections. Polysulphone is still currently the material highest rated on the UL classification for thermal endurance. The outstanding stability to hydrolysis of the polyphenylene oxide/toughened polystyrene blend (Noryl) and its excellent flow characteristics (easy processability) should also be mentioned. Another feature is their inherent flame retardancy.

The main disadvantageous characteristics of the polyarylates are their susceptibility to a number of solvents (notably chlorinated hydrocarbons and ketones), and to stress cracking. The wear characteristics are also comparatively poor and the abrasive resistance is not outstanding.

A blend of polysulphone with acrylonitrile butadiene styrene (ABS) is also available commercially under the trade name Arylon (Uniroyal). The material is, in general, similar in properties to bisphenol A polysulphone: the main purpose of introducing ABS is to improve the processing and the impact strength but, perhaps surprisingly, the processing improvement is not evident under all conditions.

3.2.6 Styrenics

The classification under this heading of polystyrene, styrene acrylonitrile copolymer and acrylonitrile butadiene styrene terpolymer is again not

TABLE 3.2(e)
Polyarylates—General Data

Polymer	Chemical structure details	Fine structure	Appearance	Melting point	Examples of commercial names
Bisphenol A polycarbonate	$\left[-O-C_6H_4-C(CH_3)_2-C_6H_4-O-CO-\right]$	Normally amorphous in mouldings (some long-range order may be present). Can be crystallised (up to about 30% crystalline content) by heating and/or solvents	Clear; slight yellowish tinge	Tc = 230–250°C Glass temperature about 149°C	Makrolon (Bayer) Lexan (General Electric) Merlon (Mobay)
Bisphenol A polysulphone	$\left[-O-C_6H_4-C(CH_3)_2-C_6H_4-O-C_6H_4-SO_2-C_6H_4-\right]$	Normally amorphous (some long-range order possible)	Clear; slight yellowish tinge	Glass temperature about 190°C	Udel (Union Carbide—USA, BXL—UK)
Polyphenylene oxide (2:6-dimethyl polyphenylene oxide)	$\left[-C_6H_2(CH_3)_2-O-\right]$	Amorphous when quenched, can be partly crystallised by heat and/or solvents	Transparent to translucent	Glass temperature about 210°C Tc = 217–257°C	
	PPO grades modified with toughened polystyrene are available (Noryl)	Amorphous	Translucent		Noryl (General Electric)

strictly chemically systematic but is technologically justifiable. As with the polyolefins, the 'engineering' properties of these polymers are not good enough for this type of application. They are, with the partial exception of the ABS terpolymer, normally strictly commodity materials. However, again in analogy with the polyolefins, incorporation of reinforcement elevates the properties sufficiently to permit engineering applications. In commodity applications the three materials share with the polyolefins the advantage of comparatively low cost. Their processability is also excellent. The other principal advantages offered by the individual materials are the excellent clarity of polystyrene and good clarity of styrene acrylonitrile, the comparatively good solvent resistance of the latter and the toughness of ABS.

The main disadvantage, shared by all three materials, is a low softening temperature. Whilst the solvent resistance of styrene acrylonitrile is the best within the group, all three materials are susceptible to solvent attack. The strength of polystyrene and ABS is comparatively low and the brittleness of polystyrene is also a well-known disadvantage.

3.2.7 Fluorocarbon Polymers

Polytetrafluoroethylene (PTFE), the oldest-established and—in some specialised applications—still the most widely used fluorocarbon polymer, does not behave as a thermoplastic in processing because its melt viscosity is so high that it does not fit into the range involved in the common methods for processing thermoplastics. Its melting point is also very high (327°C). It has also been shown that when first melted the melting temperature is higher still (342°C) and it is only subsequently that, as a result of a morphological rearrangement, the polymer will melt at the lower temperature.[2] PTFE is not injection-moulded or extruded in the conventional way and thermal conversion involves special method modifications relying mainly on sintering rather than true homogeneous melting of the polymer. For this reason the properties of PTFE are not included in Table 3.2(g) and Table 3.3(g) in Section 3.3, although a number of fillers, e.g. glass fibre, powdered petroleum coke have been used in PTFE for some time and special filled grades are also made.[2,3] These compounds are important in their particular applications, but the volume of PTFE base polymer converted to the reinforced forms is, in absolute terms, comparatively small; the high price of the polymer, as well as the high cost and difficulty of processing, are contributory factors to this situation. The very high melting point of PTFE, which is one of the causes of difficulties in processing, is a considerable service advantage, as is the stability and usability of the polymer at low temperatures. The working temperature range is normally quoted as −200°C to +250°C. Other useful properties of PTFE are its exceptional chemical resistance,

TABLE 3.2(f)
Styrenics—General Data

Polymer	Chemical structure details	Fine structure	Appearance	Melting point (Tc) and/or softening point (Ts)	Examples of commercial names
Polystyrene	$-[CH_2-CH(C_6H_5)]-$	Normally amorphous (isotactic, partly crystalline form can be prepared). High-impact ('toughened') grades are two-phase compositions with a rubber as second component	Clear, transparent (toughened grades translucent)	$Ts = 82–103°C$	Fosterene (Foster Grant) Carinex (Shell)
SAN (styrene/acrylonitrile copolymers)	$-[CH_2-CH(C_6H_5)]-$ $-[CH_2-CH(CN)]-$	Amorphous (can have 'domain' microstructure)	Clear, transparent	$Ts = 85–103°C$	Tyril (Dow Chemicals) Lustran (Monsanto)
ABS (acrylonitrile/butadiene/styrene copolymers and blends)	$-[CH_2-CH(CN)]-$ $-[CH_2-CH=CH-CH_2]-$ $-[CH_2-CH(C_6H_5)]-$	Amorphous	Translucent to opaque	$Ts = 85°C$	Terluran (BASF) Cycolac (Borg Warner Chemical)

TABLE 3.2(g)
Ethylene/Tetrafluoroethylene Copolymer (Tefzel): General Data

Chemical structure	Fine structure	Appearance	Melting point (Tc)	Examples of commercial grades
$\left[CH_2—CH_2\right]\left[CF_2—CF_2\right]$	Partly crystalline	Translucent	$Tc = 270°C$	Tefzel 200, Tefzel 280 (DuPont)

excellent electrical properties, its low coefficient of friction and the abhesive property (low surface energy).

The principal disadvantageous characteristics of PTFE are its comparatively low strength and stiffness. The wear resistance is also poor. Whilst the incorporation of fillers does not normally elevate PTFE to the position of an engineering polymer, improvements in wear resistance of about three orders of magnitude are possible.

Injection-mouldable fluorocarbon polymers have been available since the early 1960s: examples are chlorotrifluoroethylene polymer (Kel-F), polyvinylidenefluoride, fluorinated ethylene/propylene copolymer (Teflon FEP). Reinforcing fillers have been included in these materials, but the resulting property improvements have not been very significant. It is only with the reinforcement of ethylene/tetrafluoroethylene copolymer (ETFE) that the first true reinforced thermoplastic fluoro-polymer has appeared on the scene. The comonomer content in this copolymer is sufficiently high, at about 25%, to make the injection moulding easy and rapid. It also produces an improvement (in comparison with PTFE) in hardness, strength, stiffness and abrasion resistance. However, it is still low enough to preserve the characteristic properties of fluoro-polymers, in particular chemical resistance and good electrical performance although the continuous service temperature is comparatively low at about 150°C. This can be increased by radiation cross-linking to about 180°C and, in some glass-filled grades, to about 200°C. ETFE, as represented by the DuPont Tefzel resin, is the base resin in what is currently the most popular and typical reinforced thermoplastic fluorocarbon of commerce, and it is therefore the properties of Tefzel that are given in Tables 3.2(g) and 3.3(g). Another recent injection-mouldable fluoro-polymer—perfluoroalkoxy polymer (PFA)—is even closer to PTFE in its useful properties, whilst it can be melt processed in the normal way. In some respects it surpasses PTFE, for example in strength and stiffness. Its strength properties are preserved up to about 280°C.

3.2.8 Polyvinyl Chloride

Due to the scope for wide variation of its physical properties afforded by

TABLE 3.2(h)
Polyvinyl Chloride—General Data

Polymer	Chemical structure details	Fine structure	Appearance	Melting point (Tc) and/or softening point (Ts)	Examples of commercial names and grades
Polyvinyl chloride (PVC). Data for unplasticised polymer. Because of the low softening point plasticised PVC grades and vinyl copolymers are not normally reinforced	$\left[-CH_2-\underset{\underset{Cl}{\vert}}{CH}- \right]$	Commercial polymer substantially amorphous (a crystalline form has been prepared). Linear chains with some branching	Clear, transparent	$Ts = 82°C$ Glass temperature about 80°C (up to 110°C for the crystalline polymer) Tc = about 273°C (crystalline polymer)	Corvic (ICI) Bakelite PVC (Union Carbide) Vestolit (Chemische Werke Hüls) Pevikon (Fosfatbolaget AB) Kanevinyl (Kanegafuchi Chemical Industry Co.) SCON (Vinatex Ltd.) Ethyl 7042 (Ethyl Corp.)

plasticisation, the applications of PVC are particularly varied, the application range being wider than for any other thermoplastic. However, only the rigid or semi-rigid forms are of interest in engineering applications and it is rigid PVC which is mainly used as base polymer for the reinforced versions. Reinforcement can increase by a factor of three or more the strength and stiffness. It is disappointing that it does not increase the heat distortion temperature which is low. The main advantageous properties of rigid PVC are its good resistance to acids, alkalis, oils, many corrosive inorganic chemicals, oxygen, and ozone, its good water barrier properties, good electrical insulation properties, non-flammability and low creep at room temperature (for a thermoplastic material). The main limitations are the already-mentioned low softening temperature, tendency to degradation at elevated temperatures and susceptibility to attack by ketones, some chlorinated and aromatic hydrocarbons, esters, some aromatic ethers and amines and nitro compounds. Apart from the production of pipe fittings (to go with extruded PVC pipe) comparatively little PVC is injection-moulded. Many moulders still shun the material because of the comparative difficulty of processing due to the propensity to thermal decomposition. Short asbestos (chrysotile) fibres are used on a large scale as filler in PVC floor covering (tile) compositions. In these the resin is usually a vinyl chloride/acetate copolymer and the role of the asbestos is mainly that of functional filler and extender rather than reinforcement, although some improvement in mechanical properties does result.

3.2.9 Miscellaneous

3.2.9.1 Thermoplastic Polyurethane

The outstanding properties of this material are its toughness (impact resistance) and abrasion resistance. In reinforced form these are in the composite enhanced to levels higher than in any other reinforced thermoplastic. These materials would not be used where good rigidity and high strength are required.

3.2.9.2 Chlorinated Polyether (Penton)

This material was made in reinforced form but the base polymer is no longer available commercially since Hercules have stopped its production. Therefore, no data are included in the tables of this section or Section 3.3.

3.2.9.3 Other Materials

A proprietary thermoplastic polymer (Arylon-Uniroyal) described as 'polyarylether' has been produced in glass-filled form. This material is believed to be a blend of polysulphone and ABS. The main claims made for the material are that in comparison with polysulphone it has improved

TABLE 3.2(i)
Thermoplastic Polyurethane—General Properties

Chemical structure details	Fine structure	Appearance	Melting point (Tc) and/or softening point (Ts)	Examples of commercial names
Essentially linear polyurethane, usually with some terminal isocyanate groups	Amorphous	Transparent to opaque	Softening points vary with type and grade	Texin (Mobay Chemical Co.) Estane (B.F. Goodrich Chemical Co.)

flow in injection-moulding and better impact resistance. The volume used and general applicability of Arylon are too low to justify its inclusion in the tables of the present section and Section 3.3.

Injection-mouldable polyimides have from time to time been made in glass-reinforced form, but in view of their low production volumes and high cost their commercial significance is negligible.

3.3 PROPERTIES OF THE BASE POLYMERS

Tables 3.3(a)–3.3(i) list those of the important short-term mechanical properties of base polymers which are the most markedly modified by reinforcement, together with some physical and moulding properties.

The value of the short-term data is principally in direct initial comparisons between materials, either unreinforced grades against reinforced in the same material or two different materials reinforced or unreinforced one against another. For engineering design, data are also necessary on the long-term behaviour of materials under load (creep behaviour) and the effect on this of temperature and environmental factors. Detailed data on the long-term behaviour of reinforced thermoplastics are available from the commercial suppliers. Because of the sheer volume it is not practicable to include such data in the present chapter. The general roles of the matrix and reinforcement materials in determining the long-term properties of reinforced thermoplastics are discussed in Chapter 1. It is perhaps worth mentioning again that quite apart from 'normal' behaviour effects at elevated temperatures or over long times, changes in performance may take place because of degradation effects, thermal or environmental. It is for this reason that service environments should always be carefully considered in design with reinforced thermoplastics. The long-range thermal effects are reasonably well reflected in the Underwriters' Laboratories temperature index ratings (*cf.* Chapter 6).

TABLE 3.3(a)

Polyamides (Nylons): Main Physical Properties and Moulding Characteristics

Polymer	Property	Method of determination and units	Typical numerical value or description	
			Without reinforcement	Reinforced (30% glass fibre)
Nylon 6.6 (dry, as moulded)	Specific gravity	ASTM D-792	1·13–1·15	1·37
	Tensile strength	ASTM D-638 (psi)	11 800	26 000
	Tensile elongation	ASTM D-638 (%)	60	3–4
	Flexural strength	ASTM D-790 (psi)	15 000	38 000
	Flexural modulus	ASTM D-790 (psi)	410 000	1 300 000
	Compressive strength	ASTM D-695 (psi)	4 900[a]	24 000
	Izod impact strength notched/unnotched ($\frac{1}{4}$ in × $\frac{1}{2}$ in bar)	ASTM D-256 (ft lb in^{-1})	0·9/—	2·0/17
	Coefficient of linear thermal expansion	ASTM D-696 (in in^{-1} °F^{-1})	$4{\cdot}5 \times 10^{-5}$	$1{\cdot}8 \times 10^{-5}$
	Mould shrinkage ($\frac{1}{4}$ in average section)	ASTM D-955 (in in^{-1})	0·018	0·0055
	Moulding temperature injection/compression	— (°C)	260–300/260–270	265–300/260–280
	Moulding pressure injection/compression	— (psi)	Highest practicable/ 2 000–5 000	Highest practicable/ 3 000–5 000
	Moulding qualities	— —	Very good	Very good
	Distortion temperature at 264 psi	ASTM D-648 (°C)	70	255

Table 3.3(a)—*continued*

Polymer	Property	Method of determination and units	Typical numerical value or description	
			Without reinforcement	Reinforced (30% glass fibre)
Nylon 6 (dry, as moulded)	Specific gravity	ASTM D-792	1·13–1·15	1·37
	Tensile strength	ASTM D-638 (psi)	11 800	23 000
	Tensile elongation	ASTM D-638 (%)	80	3–4
	Flexural strength	ASTM D-790 (psi)	15 000	34 000
	Flexural modulus	ASTM D-790 (psi)	400 000	1 200 000
	Compressive strength	ASTM D-695 (psi)	8 800	23 000
	Izod impact strength notched/unnotched ($\frac{1}{4}$ in × $\frac{1}{2}$ in bar)	ASTM D-256 (ft lb in^{-1})	1·0/no break	2·3/20
	Coefficient of linear thermal expansion	ASTM D-696 (in in^{-1} °F^{-1})	$4·6 \times 10^{-5}$	$1·7 \times 10^{-5}$
	Mould shrinkage ($\frac{1}{4}$ in average section)	ASTM D-955 (in in^{-1})	0·016	0·0045
	Moulding temperature injection/compression	— (°C)	230–290/220–230	250–290/220–240
	Moulding pressure injection/compression	— (psi)	Highest practicable/ 2 000–5 000	Highest practicable/ 3 000–5 000
	Moulding qualities	— —	Very good	Very good
	Distortion temperature at 264 psi	ASTM D-648 (°C)	70	215

Table 3.3(a)—*continued*

Polymer	Property	Method of determination and units	Typical numerical value or description: Without Reinforcement	Typical numerical value or description: Reinforced (30% glass fibre)
Nylon 6.10 (dry, as moulded)	Specific gravity	ASTM D-792	1·07–1·09	1·30
	Tensile strength	ASTM D-638 (psi)	8 500	21 000
	Tensile elongation	ASTM D-638 (%)	85	3–4
	Flexural strength	ASTM D-790 (psi)	12 000	32 000
	Flexural modulus	ASTM D-790 (psi)	280 000	1 100 000
	Compressive strength	ASTM D-695 (psi)	3 000[a]	20 000
	Izod impact strength notched/unnotched ($\frac{1}{4}$ in × $\frac{1}{2}$ in bar)	ASTM D-256 (ft lb in^{-1})	0·6/—	2·4/20
	Coefficient of linear thermal expansion	ASTM D-696 (in in^{-1} $°F^{-1}$)	$5·0 \times 10^{-5}$	$1·5 \times 10^{-5}$
	Mould shrinkage ($\frac{1}{4}$ in average section)	ASTM D-955 (in in^{-1})	0·016	0·0045
	Moulding temperature injection/compression	— (°C)	230–270/220–230	250–290/220–240
	Moulding pressure injection/compression	— (psi)	≥ 20 000/2 000–5 000	≥ 20 000/3 000–5 000
	Moulding qualities	— —	Very good	Very good
	Distortion temperature at 264 psi	ASTM D-648 (°C)	57	215

Table 3.3(a)—*continued*

Polymer	Property	Method of determination and units	Typical numerical value or description	
			Without reinforcement	Reinforced (30% glass fibre)
Nylon 6.12 (dry, as moulded)	Specific gravity	ASTM D-792	1·05–1·08	1·30–1·37
	Tensile strength	ASTM D-638 (psi)	8 600	20 000–21 000
	Tensile elongation	ASTM D-638 (%)	100–340	4·6–4·8
	Flexural strength	ASTM D-790 (psi)	11 000–12 000	21 000–22 000
	Flexural modulus	ASTM D-790 (psi)	200 000–290 000	1 000 000–1 100 000
	Compressive strength	ASTM D-695 (psi)	2 400[a]	21 000
	Izod impact strength notched/unnotched ($\frac{1}{4}$ in × $\frac{1}{2}$ in bar)	ASTM D-256 (ft lb in^{-1})	0·85–1/—	3·0/—
	Coefficient of linear thermal expansion	ASTM D-696 (in in^{-1} °F^{-1})	$5{\cdot}0 \times 10^{-5}$	$1{\cdot}4 \times 10^{-5}$
	Mould shrinkage ($\frac{1}{4}$ in average section)	ASTM D-955 (in in^{-1})	0·011	0·002–0·004
	Moulding temperature injection/compression	— (°C)	235–250/205–215	250–290/205–225
	Moulding pressure injection/compression	— (psi)	≥ 20 000/2 000–5 000	Highest practicable/ 3 000–5 000
	Moulding qualities	— —	Very good	Very good
	Distortion temperature at 264 psi	ASTM D-648 (°C)	65	210

Table 3.3(a)—*continued*

Polymer	Property	Method of determination and units	Typical numerical value or description: Without reinforcement	Typical numerical value or description: Reinforced (30% glass fibre)
Nylon 11 (25°C and 65% RH)	Specific gravity	ASTM D-792	1·04–1·05	1·26
	Tensile strength	ASTM D-638 (psi)	7 000–8 000	13 000–14 000
	Tensile elongation	ASTM D-638 (%)	300–330	5
	Flexural strength	ASTM D-790 (psi)	—	—
	Flexural modulus	ASTM D-790 (psi)	140 000–170 000	450 000–460 000
	Compressive strength	ASTM D-695 (psi)	7 000–8 000	12 000–13 000
	Izod impact strength notched/unnotched ($\frac{1}{2}$ in × $\frac{1}{2}$ in bar)	ASTM D-256 (ft lb in^{-1})	1·8/—	—
	Coefficient of linear thermal expansion	ASTM D-696 (in in^{-1} $°F^{-1}$)	15×10^{-5}	3×10^{-5}
	Mould shrinkage ($\frac{1}{4}$ in average section)	ASTM D-955 (in in^{-1})	0·012	—
	Moulding temperature injection/compression	— (°C)	200–270/185–200	200–280/190–210
	Moulding pressure injection/compression	— (psi)	8 500–17 000/2 000–5 000	≥20 000/3 000–5 000
	Moulding qualities	— —	Very good	Very good
	Distortion temperature at 264 psi	ASTM D-648 (°C)	55	173

Table 3.3(a)—*continued*

Polymer	Property	Method of determination and units	Typical numerical value or description	
			Without reinforcement	Reinforced (30% glass fibre)
Nylon 12	Specific gravity	ASTM D-792	1·01–1·02	1·22–1·23
	Tensile strength	ASTM D-638 (psi)	8 000–9 000	17 000–18 000
	Tensile elongation	ASTM D-638 (%)	250–300	3
	Flexural strength	DIN 53452 (15 mm deflection) (psi)	9 250[b]	19 900[c]
	Flexural modulus	ASTM D-790 (psi)	165 000	1 000 000
	Compressive strength	ASTM D-695 (psi)	—	—
	Izod impact strength notched ($\frac{1}{2}$ in × $\frac{1}{2}$ in bar)	ASTM D-256 (ft lb in^{-1})	2·0–5·5	2·5–3·0
	Coefficient of linear thermal expansion	ASTM D-696 (in in^{-1} °F^{-1})	$5{\cdot}5 \times 10^{-5}$	$3{\cdot}7–4{\cdot}2 \times 10^{-5}$
	Mould shrinkage ($\frac{1}{4}$ in average section)	ASTM D-955 (in in^{-1})	0·003–0·015	0·003
	Moulding temperature injection/compression	— (°C)	190–260/175–190	220–270/180–200
	Moulding pressure injection/compression	— (psi)	8 500–14 000/2 000–5 000	≥ 20 000/3 000–5 000
	Moulding qualities	— —	Very good	Very good
	Distortion temperature at 264 psi	ASTM D-648 (°C)	55	170

[a] Stress at 1% deformation.
[b] Grilamid L209.
[c] Grilamid LV-3H.

TABLE 3.3(b)
Polyacetal: Main Physical Properties and Moulding Characteristics

Polymer	Property	Method of determination and units	Typical numerical value or description: Without reinforcement	Typical numerical value or description: Reinforced (30% glass fibre)
Homopolymer	Specific gravity	ASTM D-792	1·42	1·63
	Tensile strength	ASTM D-638 (psi)	10 000	13 000
	Tensile elongation	ASTM D-638 (%)	25–75	2
	Flexural strength	ASTM D-790 (psi)	14 100	17 500
	Flexural modulus	ASTM D-790 (psi)	410 000	1 300 000
	Compressive strength	ASTM D-695 (psi)	18 000	18 000
	Izod impact strength notched/unnotched ($\frac{1}{4}$ in × $\frac{1}{2}$ in bar)	ASTM D-256 (ft lb in^{-1})	1·4–2·3/about 20	0·8/3–4
	Coefficient of linear thermal expansion	ASTM D-696 (in in^{-1} $°F^{-1}$)	$4·5 \times 10^{-5}$	$2·4 \times 10^{-5}$
	Mould shrinkage ($\frac{1}{4}$ in average section)	ASTM D-955 (in in^{-1})	0·020–0·025	0·005
	Moulding temperature injection/compression	— (°C)	190–220/180–200	190–215/180–210
	Moulding pressure injection/compression	— (psi)	10 000–20 000/ 2 000–3 000	10 000–20 000/ 2 000–4 000
	Moulding qualities	— —	Very good	Very good
	Distortion temperature at 264 psi	ASTM D-648 (°C)	125	163

Table 3.3(b)—*continued*

Polymer	Property	Method of determination and units	Typical numerical value or description	
			Without reinforcement	Reinforced (30% glass fibre)
Copolymer	Specific gravity	ASTM D-792	1·41	1·63
	Tensile strength	ASTM D-638 (psi)	8 800	12 000–13 000
	Tensile elongation	ASTM D-638 (%)	60–75	3–4
	Flexural strength	ASTM D-790 (psi)	13 000	26 000
	Flexural modulus	ASTM D-790 (psi)	375 000	1 100 000
	Compressive strength	ASTM D-695 (psi)	16 000	—
	Izod impact strength notched/unnotched ($\frac{1}{4}$ in × $\frac{1}{2}$ in bar)	ASTM D-256 (ft lb in^{-1})	1·1/no break	1·5/3–4
	Coefficient of linear thermal expansion	ASTM D-696 (in in^{-1} °F^{-1})	$4{\cdot}7 \times 10^{-5}$	$2{\cdot}4 \times 10^{-5}$
	Mould shrinkage ($\frac{1}{4}$ in average section)	ASTM D-955 (in in^{-1})	about 0·020	0·002–0·005
	Moulding temperature injection/compression	— (°C)	180–240[a]/170–205	180–240[a]/170–205
	Moulding pressure injection/compression	— (psi)	10 000–20 000/ 1 000–5 000	10 000–20 000/ 2 000–4 000
	Moulding qualities	— —	Very good	Very good
	Distortion temperature at 264 psi	ASTM D-648 (°C)	110	157

[a] Above 220°C residence times must be short.

TABLE 3.3(*c*)

Polyolefins: Main Physical Properties and Moulding Characteristics

Property	Method of determination and units	Typical values or descriptions							
		HD Polyethylene		Polypropylene		Poly-4-methylpentene-1		Polybutene	
		Unreinforced	30% glass fibre	Unreinforced	30% glass fibre	Unreinforced	30% glass fibre[a]	Unreinforced	30% glass fibre[b]
Specific gravity	ASTM D-792	0·941–0·965	1·17	0·90–0·91	1·13	0·83	0·96	0·91	1·13
Tensile strength	ASTM D-638 (psi)	3 000–5 000	10 000	5 000	9 100	4 000	7 200	2 200	4 600
Tensile elongation	ASTM D-638 (%)	20–1 000	2–3	200–700	2–3	15	4–5	350	3–4
Flexural strength	ASTM D-790 (psi)	3 000	11 500	7 000	11 000	—	10 000	—	7 200
Flexural modulus	ASTM D-790 (psi)	100 000–200 000	900 000	170 000–250 000	850 000	210 000	600 000	26 000	450 000
Compressive strength	ASTM D-695 (psi)	2 700–3 600	—	5 500–7 000	8 500	—	—	—	—
Izod impact strength notched/unnotched	ASTM D-256 (ft lb in^{-1})	0·5–20/— (½ × ½ in bar)	1·1/8·9 (¼ × ½ in bar)	0·5–2/— (½ × ½ in bar)	1·6/5–6 (¼ × ½ in bar)	0·8/— (½ × ½ in bar)	2·9/4–5 (¼ × ½ in bar)	15/no break (¼ × ½ in bar)	5·9/10–11 (¼ × ½ in bar)
Coefficient of linear thermal expansion	ASTM D-696 (in in^{-1} °F^{-1})	$6·1 \times 10^{-5}$–$7·2 \times 10^{-5}$	$2·7 \times 10^{-5}$	$3·8 \times 10^{-5}$–$5·8 \times 10^{-5}$	$2·0 \times 10^{-5}$	$6·5 \times 10^{-5}$	—	—	—
Mould shrinkage (¼ in average section)	ASTM D-955 (in in^{-1})	0·02–0·05	0·0025	0·01–0·025	0·002–0·008	0·015–0·030	—	—	—
Moulding temperature injection/compression	— (°C)	200–280/ 150–230	205–287/ 205–250	205–280/ 170–280	230–290/ 190–280	260–300/ 250–280	270–300/ 280	190–290/[c] 190–250	
Moulding pressure injection/compression	— (psi)	10 000–20 000/ 500–800	up to 20 000/ 1 000–4 000	10 000–20 000/ 500–1 000	15 000–20 000/ 1 000–4 000	8 000–20 000/ 1 000–3 000	10 000–20 000/ 1 000–3 000	10 000–30 000/ 500–1 000	15 000–30 000/ 500–1 000
Moulding qualities	— —	Very good	Very good	Very good	Very good	Good	Good	Good	Good
Distortion temperature at 264 psi	ASTM D-648 (°C)	43–55	127	55–60	146	60	115	57	113

[a] Thermocomp LF(LNP).
[b] Thermocomp EF(LNP).
[c] Vestolen BT(Hüls).

TABLE 3.3(d)

Thermoplastic Polyesters: Main Physical Properties and Moulding Characteristics

Polymer	Property	Method of determination and units	Typical numerical value or description	
			Without reinforcement	Reinforced (30% glass fibre)
Polybutylene terephthalate (polytetra-methylene terephthalate)	Specific gravity	ASTM D-792	1·31–1·32	1·59
	Tensile strength	ASTM D-638 (psi)	8 200	19 000–19 500
	Tensile elongation	ASTM D-638 (%)	150–300	2–4
	Flexural strength	ASTM D-790 (psi)	12 000–13 000	27 000
	Flexural modulus	ASTM D-790 (psi)	330 000–340 000	1 200 000
	Compressive strength	ASTM D-695 (psi)	12 000–13 000	17 000–18 000
	Izod impact strength notched/unnotched ($\frac{1}{4}$ in × $\frac{1}{2}$ in bar)	ASTM D-256 (ft lb in^{-1})	1·0/—	1·6/10
	Coefficient of linear thermal expansion	ASTM D-696 (in in^{-1} $°F^{-1}$)	$4–6 \times 10^{-5}$	$1·2–5·4 \times 10^{-5}$
	Mould shrinkage ($\frac{1}{4}$ in average section)	ASTM D-955 (in in^{-1})	0·017–0·023	0·002–0·006
	Moulding temperature injection/compression	— (°C)	240–250/225–250	240–250/225–250
	Moulding pressure injection/compression	— (psi)	8 000–23 000/ 1 000–4 000	8 000–23 000/ 2 000–4 000
	Moulding qualities	— —	Good	Good
	Distortion temperature at 264 psi	ASTM D-648 (°C)	55	200–212

Table 3.3(d)—*continued*

Polymer	Property	Method of determination and units	Typical numerical value or description: Without reinforcement	Reinforced (20% glass fibre)
Polyethylene terephthalate Unreinforced—Hostadur KP 4022; Reinforced (20% glass fibre)—Hostadur KVP 8022 6V1	Specific gravity	Method as shown	1·37 (buoyancy)	1·50 (DIN 53479)
	Tensile strength	DIN 53455 (kgf cm^{-2})	740	1 300–1 350
	Tensile elongation	DIN 53455 (%)	50	4
	Flexural strength	DIN 53452 (kgf cm^{-2})[a]	1 250	1 500
	Flexural modulus	—	—	—
	Compressive strength	—	—	—
	Izod impact strength notched/unnotched	DIN 53453 (kgf cm cm^{-2})	4/no break	5–6/25
	Coefficient of linear thermal expansion	— (in in^{-1} °C^{-1})	7×10^{-5}	3×10^{-5} (dilatometer)
	Mould shrinkage ($\frac{1}{4}$ in average section)	— (%)	0·8–2	0·25–0·6
	Moulding temperature injection/compression	— (°C)	260–270/260–270	260–280/260–270
	Moulding pressure injection/compression	— (psi)	Highest practicable/ 1 000–4 000	Highest practicable/ 2 000–4 000
	Moulding qualities	— —	Fairly good	Fairly good
	Distortion temperature at 264 psi	ASTM D-648 (°C)	68	221

[a] Injection moulded specimen No. 3, test speed V.

TABLE 3.3(e)
Polyarylates: Main Physical Properties and Moulding Characteristics

Property	Method of determination and units	Typical values or descriptions					
		Polycarbonate		Polysulphone		Modified polyphenylene oxide (Noryl)	
		Unreinforced	30% glass fibre	Unreinforced	30% glass fibre	Unreinforced	30% glass fibre
Specific gravity	ASTM D-792	1·20	1·43	1·24	1·45	1·06	1·27
Tensile strength	ASTM D-638 (psi)	9 500	16 000	10 200	18 000	9 500	18 500
Tensile elongation	ASTM D-638 (%)	90	4–6	50–100	3	60	3–4
Flexural strength	ASTM D-790 (psi)	13 500	28 000	12 000	24 000	13 500	23 000
Flexural modulus	ASTM D-790 (psi)	340 000	1 200 000	390 000	1 200 000	360 000	1 150 000
Compressive strength	ASTM D-695 (psi)	12 500	22 000	13 900	24 000	16 500	17 900
Izod impact strength notched/unnotched ($\frac{1}{4}$ in × $\frac{1}{2}$ in bar)	ASTM D-256 (ft lb in^{-1})	2·7/60	3·7/17	1·3/>60	1·8/14	1·2/no break	1·7/9–10
Coefficient of linear thermal expansion	ASTM D-696 (in in^{-1} $°F^{-1}$)	$3·75 \times 10^{-5}$	$1·25 \times 10^{-5}$	$3·1 \times 10^{-5}$	$1·4 \times 10^{-5}$	$3·3 \times 10^{-5}$	$1·4 \times 10^{-5}$
Mould shrinkage ($\frac{1}{4}$ in average section)	ASTM D-955 (in in^{-1})	0·007	0·002 5	0·007–0·008	0·002–0·003	0·005–0·007	0·002–0·004
Moulding temperature injection/compression	— (°C)	250–345/ 250–325	270–340/ 260–325	330–380/ 285–320	345–390/ 290–330	245–300/ 205–240	260–300/ 230–250
Moulding pressure injection/compression	— (psi)	10 000– 20 000/ 1 000–2 000	10 000– 20 000/ 2 000–4 000	15 000– 25 000/ 1 000	15 000– 25 000/ 2 000–4 000	14 000– 20 000/ 500–1 000	15 000– 40 000/ 2 000–4 000
Moulding qualities	— —	Very good	Very good	Very good	Good	Very good	Very good
Distortion temperature at 264 psi	ASTM D-648 (°C)	140	149	174	185	130	155

TABLE 3.3(f)
Styrenics: Main Physical Properties and Moulding Characteristics

Property	Method of determination and units	Typical values or description					
		Polystyrene		SAN		ABS (medium impact grade)	
		Unreinforced	30% glass fibre	Unreinforced	30% glass fibre	Unreinforced	30% glass fibre
Specific gravity	ASTM D-792	1·04	1·28	1·04–1·10	1·31	1·04–1·07	1·28
Tensile strength	ASTM D-638 (psi)	5 000–10 000	13 500	9 000–12 000	18 000	5 500–8 000	14 500
Tensile elongation	ASTM D-638 (%)	1·0–2·3	1–2	1·5–3·7	2–3	5–25	3–4
Flexural strength	ASTM D-790 (psi)	10 000–15 000	16 200	14 000–19 000	22 000	9 500–13 000	18 500
Flexural modulus	ASTM D-790 (psi)	400 000–500 000	1 200 000	≤ 550 000	1 500 000	300 000–400 000	1 100 000
Compressive strength	ASTM D-695 (psi)	11 500–16 000	18 000	14 000–17 000	21 000	8 000–12 500	14 500
Izod impact strength notched/unnotched	ASTM D-256 (ft lb in^{-1})	0·25–0·40/— ($\frac{1}{4}\times\frac{1}{2}$ in bar)	1·0/2–3 ($\frac{1}{4}\times\frac{1}{2}$ in bar)	0·3–0·5/— ($\frac{1}{2}\times\frac{1}{2}$ in bar)	1/3–4 ($\frac{1}{4}\times\frac{1}{2}$ in bar)	3–6/— ($\frac{1}{2}\times\frac{1}{2}$ in bar)	1·4/6–7 ($\frac{1}{4}\times\frac{1}{2}$ in bar)
Coefficient of linear thermal expansion	ASTM D-696 (in in^{-1} °F^{-1})	$3{\cdot}3\times10^{-5}$–$4{\cdot}8\times10^{-5}$	$1{\cdot}9\times10^{-5}$	$3{\cdot}6\times10^{-5}$–$3{\cdot}8\times10^{-5}$	$1{\cdot}8\times10^{-5}$	$4{\cdot}4\times10^{-5}$–$5{\cdot}5\times10^{-5}$	$1{\cdot}6\times10^{-5}$
Mould shrinkage ($\frac{1}{4}$ in average section)	ASTM D-955 (in in^{-1})	0·002–0·006	0·001	0·002–0·007	0·001	0·005–0·008	0·0015
Moulding temperature injection/compression	— (°C)	160–260/ 130–205	240–290/ 200–240	190–270/ 150–205	205–290/ 190–240	220–260/ 150–230	230–275/ 220–240
Moulding pressure injection/compression	— (psi)	10 000–25 000/ 1 000–2 000	15 000–40 000/ 2 000–4 000	10 000–33 000/ 1 000–2 000	15 000–40 000/ 2 000–5 000	3 000–25 000/ 1 000–2 000	15 000–40 000/ 2 000–5 000
Moulding qualities	— —	Very good	Very good	Good	Good	Good	Fair
Distortion temperature at 264 psi	ASTM D-648 (°C)	≤ 105	≤ 105	88–105	102	85–105	105

TABLE 3.3(g)

Ethylene/Tetrafluoroethylene Copolymer (Tefzel): Main Properties and Moulding Characteristics

Property	Method of determination and units	Typical numerical value or description	
		Without reinforcement (Tefzel 200)	Reinforced (25% glass fibre —Tefzel 706-25)
Specific gravity	ASTM D-792	1·70	1·86
Tensile strength	ASTM D-638 (psi)	6 500	12 000
Tensile elongation	ASTM D-638 (%)	100–300	8
Flexural strength	ASTM D-790 (psi)	10 000[a]	19 000[a]
Flexural modulus	ASTM D-790 (psi)	200 000	950 000
Compressive strength	ASTM D-695 (psi)	7 100	10 000
Izod impact strength notched/unnotched ($\frac{1}{4}$ in × $\frac{1}{4}$ in bar)	ASTM D-256 (ft lb in^{-1})	No break/No break[a]	7–8/17–18[a]
Coefficient of linear thermal expansion	ASTM D-696 (in in^{-1} $°F^{-1}$)	9×10^{-5}	3×10^{-5}
Mould shrinkage ($\frac{1}{8}$ in average section)	ASTM D-955 (in in^{-1})	0·015–0·020 in flow direction 0·035–0·045 in transverse direction	0·002–0·003 in flow direction About 0·030 in transverse direction
Injection moulding temperature	— (°C)	295–340	295–340
Injection moulding pressure	— (psi)	3 000–15 000	3 000–15 000
Moulding qualities	— —	Good	Good
Distortion temperature at 264 psi	ASTM D-648 (°C)	74	210

[a] 30%-filled material: Thermocomp UF 1006 (LNP).

TABLE 3.3(h)
Polyvinyl Chloride: Main Physical Properties and Moulding Characteristics

Property	Method of determination and units	Typical numerical value or description	
		Without reinforcement	Reinforced (25% glass fibre)
Specific gravity	ASTM D-792	1·33–1·45	1·53
Tensile strength	ASTM D-638 (psi)	5 000–9 000	16 000
Tensile elongation	ASTM D-638 (%)	1–25	2–3[a]
Flexural strength	ASTM D-790 (psi)	10 000–16 000	18 000[a]
Flexural modulus	ASTM D-790 (psi)	300 000–500 000	1 250 000
Compressive strength	ASTM D-695 (psi)	7 000–13 000	12 000[a]
Izod impact strength notched/unnotched	ASTM D-256 (ft lb in^{-1})	0·4–20·0/— ($\frac{1}{2} \times \frac{1}{2}$ in bar)	1·1/7–8 ($\frac{1}{4} \times \frac{1}{2}$ in bar)
Coefficient of linear thermal expansion	ASTM D-696 (in in^{-1} °F^{-1})	$2{\cdot}8 \times 10^{-5}$–10×10^{-5}	$1{\cdot}5 \times 10^{-5}$
Mould shrinkage ($\frac{1}{4}$ in average section)	ASTM D-955 (in in^{-1})	0·001–0·004	0·000 7–0·001
Moulding temperature injection/compression	— (°C)	150–185/140–185	175–195/160–180
Moulding pressure injection/compression	— (psi)	20 000–40 000/1 000–2 000	≥ 20 000/2 000–5 000
Moulding qualities	— —	Fair–Good	Fair
Distortion temperature at 264 psi	ASTM D-648 (°C)	60–80	71

[a] 15% glass fibre.

TABLE 3.3(1)

Thermoplastic Polyurethanes—Main Physical Properties and Moulding Characteristics

Property	Method of determination and units	Typical numerical value or description		
		Without reinforcement (Texin 355D : Goodrich)	Reinforced with	
			30% glass fibre (Thermocomp TF : LNP)	40% 'long' glass fibre (Urafil G100-40 : Fiberfil)
Specific gravity	ASTM D-792	1·23	1·46	1·53
Tensile strength	ASTM D-412 (psi)	6 200 (ultimate)		
Tensile strength	ASTM D-638 (psi)		8 200 (yield)	10 000 (yield)
Tensile elongation	ASTM D-412 (%)	480 (ultimate)		
Tensile elongation	ASTM D-638 (%)		—	2 (yield)
Flexural strength	ASTM D-790 (psi)	—	—	7 000
Flexural modulus	ASTM D-790 (psi)	—	200 000	360 000
Compressive strength	ASTM D-575 (psi)	2 250 (stress at 25% strain)		
Compressive strength	ASTM D-695 (psi)		—	7 000 (yield)
Notched Izod impact strength ($\frac{1}{4}$ in × $\frac{1}{2}$ in bar)	ASTM D-256 (ft lb in^{-1})	—	9·5	10
Coefficient of linear thermal expansion	ASTM D-696 (in in^{-1} $°F^{-1}$)	—	$2{\cdot}5 \times 10^{-5}$	$1{\cdot}1 \times 10^{-5}$
Mould shrinkage ($\frac{1}{4}$ in average section)	ASTM D-955 (in in^{-1})	0·010–0·015	—	0·001
Injection moulding temperature	— (°C)	200–220 (melt)	180–220	180–220
Injection moulding pressure	— (psi)	8 000–15 000	8 000–20 000	
Moulding qualities	— —	Good	Good	Good
Distortion temperature at 264 psi	ASTM D-648 (°C)	—	76	—

REFERENCES

1. 'Kematal', ICI Technical Data Sheet K.TD 16, January 1971.
2. Wells, D. M. (1970). Paper presented at the Plastics Institute Conference on Reinforced Thermoplastics, October, Solihull, England.
3. 'Thermocomp' Technical Literature: Liquid Nitrogen Processing Corp. and Technical Literature of companies listed in the Tables of Chapter 2.

BIBLIOGRAPHY

Anon. Melt Processable Fluorocarbons, *Europlastics* 1973, **46,** 11, 65.

Davis, J. M. (1970). 'Properties and Applications of Glass-filled PPO.' Paper presented at the Plastics Institute Conference on Reinforced Thermoplastics, October, Solihull, England.

Bernardo, A. C. (1970). How to Get More From Glass-fibre Reinforced HDPE, *SPE Journal,* **26,** 39.

CHAPTER 4

REINFORCEMENTS

4.1 GENERAL FEATURES

By definition, the main function of a reinforcing filler is to improve the mechanical properties of the base polymer. Of these stiffness and strength, in that order, are the most important among 'short-term' properties in engineering applications. Resistance to creep and fatigue failure are the principal long-term properties. The effect of heat on both these groups of properties, as reflected in the temperature for deflection under load and maximum temperature for continuous service, is also important as is dimensional stability. All these properties can be upgraded by reinforcing fillers (see Chapters 3 and 6). In most applications the proper balance of properties is no less important than improvements in individual property values. Non-mechanical properties of the base polymer, e.g. electrical properties, abrasion resistance, flammability, may also be strongly modified by the presence of reinforcing fillers.

At a reasonable level of loading, and given good interfacial contact and adhesion between filler and matrix, improvements in stiffness and strength will be functions of these properties of the filler. With fibrous fillers the improvements can be further magnified by the influence of the fibre aspect ratio and the anisotropic effect of fibre orientation. These factors have been considered in Chapter 1 (Section 1.4).

It is for these reasons that the most effective reinforcing fillers are fibres of high modulus and strength. Glass fibres, which are non-crystalline, or asbestos—a crystalline fibre—provide the reinforcement in most commercial fibre-reinforced thermoplastics. Carbon fibres and 'whiskers' (single-crystal fibres) are the other crystalline fibres used as reinforcement. Other things being equal, fibre stiffness tends to increase with crystallinity: of the fibres just mentioned glass fibres have the lowest modulus and specific modulus (*cf.* Table 4.1).

In general, the strength of a material is determined by its chemical structure and its morphology (fine structure) which includes molecular order (the ultimate being perfect crystallinity) and orientation (of individual molecules or crystals). From relevant basic structural data the maximum theoretical strength and stiffness of a material may be worked out, with the

TABLE 4.1
Mechanical Properties of some Reinforcement Fibres

Fibre	Density ($g\ cm^{-3}$)	Tensile strength (10^5 psi)	Young's modulus (10^6 psi)	Specific modulus
Carbon, type 1 (high modulus)	2·0	3	60	30
Carbon, type 2 (high strength)	1.74	4	40	23
Whiskers				
Silicon nitride	3·2	20	57	19
Potassium titanate	3·2	>10	40	12·5
Silicon carbide	3·2	14–29	68–74	22·2
Alumina (sapphire)	4·0	41	103·5	25·9
E glass	2·55	4–5·4	10·5–11	4·2
S glass	2·48	6·5–7	12·5	5·1
Asbestos				
Chrysotile	2·55	4·4	23	9·0
Crocidolite	3·37	5	27	7·4
Anthophyllite	2·85–3·1	2–4	21	7·3

Information based in part on the data of Hollingsworth and Sims,[1] Hartley,[2] and Hodgson,[3] abstracted, with permission, respectively, of the Controller, Her Majesty's Stationery Office (Crown copyright holders), Engineering Materials & Design, IPC Industrial Press Ltd., and the Royal Institute of Chemistry.

aid of such concepts as cohesive energy density. The actual strengths and moduli of most materials in bulk are normally lower, frequently by several orders of magnitude, than the theoretically calculated maximum values. The factors responsible for this fall into two general groups: irregularity of internal structure and surface imperfections.

In non-crystalline materials the arrangement of molecules is more or less irregular, and structural voids may be present, or formed under stress. In polycrystalline materials inter-crystal boundaries may constitute lines or planes of weakness, and non-uniformity of orientation of individual crystals tends to reduce strength and stiffness. Finally, defects and dislocations within individual crystals also affect these properties adversely. Such intracrystalline imperfections are the main reason why the strength and modulus of single-crystal structures (e.g. whiskers) fall short of the theoretical. Surface imperfections and micro-cracks are the main 'external' structural factors responsible for lowering the strength and modulus of both crystalline and non-crystalline materials.

Materials of high potential strength and stiffness usually approach the maximum values of these properties more closely when in fibrous form.

This is because in fibres the molecular order is normally greater, the fine structure more uniform and the surface imperfections may be fewer than in the bulk material. The ultimate example is a perfect whisker with undamaged surface: such fibrous crystals can closely approach the maximum theoretical strength and stiffness.[4] Fibres can also be treated (sized) to reduce the formation of surface defects which arise as a fresh surface ages and/or suffers damage through handling. These effects can seriously reduce the ultimate fibre strength. For example, tensile strengths of 285 000 and 540 000 psi have been quoted[5] respectively for ordinary and 'undamaged' E glass fibre (*cf.* also Section 4.2).

The large specific surface (surface per unit volume) of fine fibres makes for large interfacial contact area with the polymer matrix which in turn increases the strength of the composite if the interfacial adhesion is good.

The fillers incorporated into commercial thermoplastics as true reinforcement, in contradistinction to extenders and pigments, are listed in Table 4.2. Some recent developments are also mentioned in Chapter 8. Particulate lubricant fillers (PTFE, MoS_2) may be incorporated along with the reinforcing fillers to reduce surface friction of the composite.

The differences between some of the most relevant properties of the reinforcements and the base polymers, which are responsible for the main reinforcing effects, are summarised in Table 4.3.

4.2 GLASS FIBRES

4.2.1 Nature and Structure of Glass

The appropriate ASTM standard* defines glass, in general terms, as:

> 'an inorganic product of fusion which has cooled to a rigid condition without crystallising'.

The last two words are particularly important in that they relate to the principal morphological characteristic of glass, *viz.* its amorphous nature. This is manifested *inter alia* in the optical, electrical and structural anisotropy of normal glass in bulk, in its typically 'amorphous' X-ray diffraction spectrum, and lack of a sharp transition point on cooling from melt to solid or fusion from solid to melt.

Chemically, normal glass is a composition (but not a definite chemical compound) of predominantly silica, with varying amounts of oxides of other elements. Fused silica and quartz are virtually pure SiO_2. Such one-component glasses can possess structural regularity in the form of network arrangements of their constituent atoms. Fused SiO_2 and quartz have so far been used only in development quantities as fibrous reinforcement for thermoplastics.

* C 162–71 Standard definitions of terms relating to glass and glass product.

TABLE 4.2
Reinforcing Fillers in Thermoplastics

Reinforcement	Nature and/or type used	Surface treatments	Base polymers in which used	Main improvements in engineering properties[a]	Examples of end use of composite
Glass fibres	Predominantly E glass	Coating with coupling agents to improve interfacial adhesion to polymer (*cf.* Section 4.2.5)	Virtually all	Greatly increased strength and stiffness, reduced shrinkage. In many cases increased deflection temperature. Improved creep resistance	Very numerous—see Chapter 8
Glass beads (Ballotini)	A glass	Coating with coupling agents to improve interfacial adhesion to polymer (*cf.* Section 4.2.5)	Nylon, ABS, HD polyethylene, polypropylene, SAN	Increased strength, stiffness, hardness and abrasion resistance. Reduced shrinkage	Camera and computer components (nylon), conveyor rollers (HDPE), bobbins (nylon), water dispensers (ABS), housewares
Asbestos fibres	Anthophyllite, Amosite, Crocidolite, Chrysotile	Not normally necessary to improve adhesion. May be applied to reduce it to increase toughness of the composite. Coatings to improve heat stability in some cases	Polypropylene Polystyrene Nylon PVC	Increased stiffness and strength. Higher deflection temperature. Improved creep resistance	Automotive parts, electrical components, pipe fittings (polypropylene), flooring (PVC)

Table 4.2—*continued*

Reinforcement	Nature and/or type used	Surface treatments	Base polymers in which used	Main improvements in engineering properties[a]	Examples of end use of composite
Carbon fibres	Type 1 (high modulus) Type 2 (high strength)	Special surface treatments for coupling to particular polymers now available	Nylon ETFE polymer Polyacetal Polycarbonate	Improved stiffness, some strength improvement, some lubricity	Gear wheels, bearings, golf-club faces (polycarbonate), tennis racquets (nylon)
Whiskers (Fybex)	Potassium titanate	Not usual	Polypropylene, ABS, nylon, polysulphone, modified PPO, PVC	Increased strength and stiffness (isotropic effect). Improved surface finish. Reduced shrinkage	Electroplated and/or thermoformed products (ABS), profiles, piping, cladding (PVC), gears, camera parts (nylon)
Talc (particles: needle or platelet shape)	Basically magnesium silicate	Coupling agents may be used to improve composite properties	Polypropylene	Increased stiffness, heat deflection temperature and hardness	Textile machinery parts, electronics, office and domestic equipment
Wollastonite (particles: needle-shaped)	Basically calcium metasilicate	Coupling agents may be used (e.g. Dow Corning XZ-8-5069)	Polypropylene	Increased stiffness, hardness and deflection temperature	Textile machinery parts, electronics, office and domestic equipment
Chalk (particles: roughly spherical)	Calcium carbonate	Stearate coating to promote dispersion in polymer. Effect of silane coupling agents negligible	Polypropylene PVC	Increased stiffness and hardness. Silky matt surface finish	Electrical and radio parts, office and domestic equipment

[a] In some cases impact strength (toughness) is adversely affected.

TABLE 4.3

Comparison of the Main Physical and Mechanical Properties of Base Polymers with those of Reinforcement Materials used in Commercial Reinforced Thermoplastics

	Properties (approximate value ranges)						
	Specific gravity (g cm^{-3})	Tensile strength (psi)	Modulus (psi)	Elongation (%)	Coefficient of linear thermal expansion (°C^{-1})	Heat distortion temperature (°C)	Creep at normal temperatures
Base polymers	0·9–1·4	$3–12 \times 10^3$	$50–500 \times 10^3$	1–1 000	$5–18 \times 10^5$	43–174	Occurs in varying degrees
Reinforcements	1·7–3·4	$2–10 \times 10^5$	$100–600 \times 10^5$	0·5–5	Approx. 10^{-6}	> 500	Virtually none
Difference (approximate factor)	× 2	× 100	× 100	Reduced × 2–200	Reduced × 100	> × 3	—
Order of maximum improvement on incorporation of reinforcement in polymer (very approximate)	—	× 3	× 7	Reduced × 50	Reduced × 50	× 1·5	Considerably reduced

TABLE 4.4

Composition and some Properties of Glass Fibres used in the Plastics Industry

Glass type	Approximate composition (wt %)					Specific gravity (g cm^{-3})	Softening temp. (°C)	Tensile strength[b] (psi)	Young's modulus (psi)	Remarks
	SiO_2	Al_2O_3 and Fe_2O_3	CaO and MgO	Na_2O and K_2O	B_2O_3 and BaO					
E	52·4–53·2	14·4–14·8	21·4–21·8	<1[a]	9·3–10·6	2·54–2·56	750–800	400–540 × 10^3	10·5–11·0 × 10^6	Resistant to water attack. Good electrical properties
S (or S-994)	64·0–64·3	25·0–26·0	10·0–10·3	0·0–0·3	<0·01	2·48–2·49	850–880	650–700 × 10^3	12·4–12·5 × 10^6	High-strength glass
C	63·6–64·6	4·0–4·1	16·6–16·7	9·1–9·6	5·6–6·7	2·48–2·49	About 690	420–540 × 10^3	10·0 × 10·5 × 10^6	Resistant to acid attack
A (Soda lime glass)	72·5	0·7–1·5	12·5–13·1	13·5–13·8	—	2·45–2·49	720–730	440–470 × 10^3	10·0–10·5 × 10^6	Used in RTP in the form of glass beads (*cf.* Section 4.3)

[a] According to the definition in BS 3691:1969, E glass is glass containing less than 1% alkali expressed as Na_2O.
[b] Freshly drawn single fibre.

4.2.2 Types of Glass Fibres

Some relevant information on the main glass fibre types is summarised in Table 4.4.

In addition to the more common glass fibres listed in Table 4.4 special high-strength glasses have also been prepared in fibrous form, reputedly based on magnesium aluminium silicate compositions. These are the R glass (Société du Verre Textile—France) and 4H1 glass (Aerojet General—USA). Their moduli as well as strengths are higher than those of E glass,[5] whilst the densities are similar.

4.2.3 Manufacture of Glass Fibres

Glass fibres used as reinforcement in thermoplastics are manufactured, in continuous filament form, by the drawing process. Other processes—the centrifugal process and the blowing (jet) process—are also available, but these produce staple fibres which are used for other purposes.[6]

The salient features of the drawing process are basically the same for all commercially available chemical types of glass.[1,6] The method is schematically illustrated in Fig. 4.1.

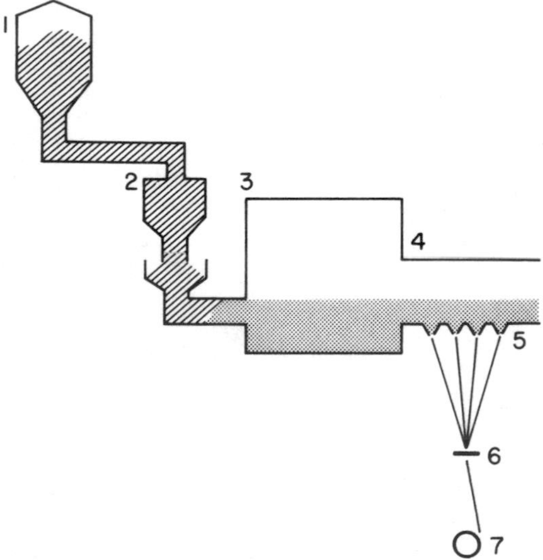

FIG. 4.1 Production of glass fibres by the continuous filament process, schematic representation. 1: Glass batch. 2: Batch charge. 3: Tank. 4: Forehearth. 5: Filament forming. 6: Gathering and sizing. 7: High-speed winder. (Courtesy of Fibreglass Ltd.)

Molten glass is passed from a furnace (the 'tank'), through a forehearth, to a series of bushings, each containing several hundred accurately dimensioned jets (forming tips).

A constant level of melt is maintained in the tank as fine filaments are drawn mechanically downwards from the bushing tips, at several thousand metres per minute. For a given type and melt viscosity of glass the filament diameter is controlled by the drawing speed. The diameters of commercial filaments lie within the range 9–18 μm, although still finer filaments (down to about 3 μm) can be produced. The filaments from each bushing pass through a collecting point where size is put on, are brought together to form a strand, and the strands are wound singly or as multiple strands on a high-speed winder.

The main functions of the size are to protect the filament surface and to hold the strand together. Coupling agents (*cf.* Section 4.2.5) are also normally applied as a component of the size, although they may be applied in a separate operation. Size may have to be removed, usually by burning off, before separate treatment with a coupling agent. This is commonly done e.g. when glass fabric is to be treated with a coupling agent. The advantages of applying surface treatments as soon as possible after forming the fibres are self-evident: the protective coating counteracts development of faults through external damage to the surface, whilst a coupling agent can be expected to react most readily and fully with the freshly formed surface.

The manufacturers of glass fibres and their trade names are listed in Table 4.5.

TABLE 4.5
Glass Fibre Producers and their Products
(Abstracted, with permission, from the *Textile Organon,* June 1974, Textile Economics Bureau, Inc.)

Country	Producer	Trade name
Argentina (Llavallol, Buenos Aires)	Vidrieria Argentina S.A. (VASA)	Texover
Australia (Dandenong, Victoria)	Australian Fibre Glass Pty. Limited	A.C.I. Fibreglass
Belgium (Battice, Liege)	Owens-Corning Fiberglas Europe S.A.	Fiberglas
Brazil (Rio Claro, Sao Paulo)	Ocfibras (formerly Owens-Corning Fiberglas)	Fiberglas
Canada (Guelph, Ontario)	Fiberglas Canada Limited	Fiberglas
Colombia (Bogota)	Owens-Corning Fiberglas Colombia, S.A.	Fiberglas

Table 4.5—*continued*

Country	Producer	Trade name
Czechoslovakia (Litomysl, Bohemia and Trnava Slovakia)	Vertex N.P. (Vertex National Corporation)	Vertex
Finland (Karhula)	Karhulan Kuitulasi OY	Kuitulasi
France (Chambery, Savoie)	Saint-Gobain, Division Verre Textile	Silionne, Stratifil, Stratimat, Verrane
(La Voulte, Ardeche)	Societe Elastover	Pilot plant
(L'Ardoise, Gard)	Societe Owens-Corning Fiberglas France, S.A.	Fiberglas
Germany, Federal Republic of (Aachen)	Gevetex-Textilglas G.m.b.H.	Gevetex
India (Thana, Nr. Bombay, Maharashtra)	Fibreglass Pilkington Limited	—
Italy (San Salvo of Vasto, Abruzzi)	Societa Italiana Vetro, S.p.A.	SIV
(Besana Brianza, Milano and Livorno, Tuscany)	Vetreria Italiana Balzaretti Modigliani S.p.A.	Filospun (spun strand and roving) TBtex (staple, yarn and sliver) Unifilo (continuous mat) Vetrotex (yarn) Vetrotex RMP (roving and mat)
Japan (Shonan and Koga)	Asahi Fiber Glass KK (Asahi Fibre Glass Co. Ltd.) Asahi PPG Ltd. (Plant site undetermined)	Glasslon
(Moka)	Fuji Fiber Glass Co.	No trade mark
(Tsu)	Nippon Garasu Sen-1 KK (The Nippon Glass Fibre Co. Ltd.)	Micro Glass
(Fujisawa, Yuki and Taru)	Nippon Muki Sen-1 Kogyo KK (The Nippon Mineral Fibre Mfg. Co.)	Superfine
(Fukushima)	Nitto Boseki KK (Nitto Boseki Co. Ltd.)	Iceberg

Table 4.5—*continued*

Country	Producer	Trade name
(Koriyama and Suzuka)	Paramount Glass Kogyo KK (Paramount Glass Mfg. Co.)	Fether Glass
(Fujisawa)	Sanriku Fiberglass Industry Co. Ltd.	Sanrifine
(Fushimi)	Unitika U.M. Glass Co. Ltd.	Unitika U.M.
Mexico (Zacatenco (D.F.))	Vitro-Fibras, S.A.	Vitro-Fibras
Netherlands (Etten)	N.V. Glaswolfabriek Isoverbel	Isoverbel
(Hoogezand–Groningen)	Silenka B.V.	Enkafort Silenka
Norway (Oslo)	Norsk Glassfiber	—
Poland (Lodz)	Zaklady Wlokien Sztucznych 'Anilana'	Polsilon
(Krosno)	Krosnienskie Huty Szkla	—
South Africa (New Era Springs, Transvaal)	Fibreglass South Africa (Pty.) Limited	—
Spain (Alcalá de Henares, Madrid)	Fibras Minerales, S.A.	Vitrotex
Sweden (Falkenberg)	Scandinavian Glasfiber Aktiebolag	—
Union of Soviet Socialist Republics	No reliable information	
UK (St. Helens, Lancs., Wrexham, Clwyd, Wales)	Fibreglass Ltd.	Fibreglass
(Sherborne, Dorsetshire)	Marglass Ltd.	Marglass
(Camberley, Surrey, Hindley Green, Lancashire and Dungannon, Tyrone, Northern Ireland)	TBA Industrial Products Ltd. (formerly Turner Bros. Asbestos Co. Ltd.)	Deeglas Duraglas
USA (Nashville, Tenn.)	Ferro Corporation, Fiber Glass Division	Roving, woven roving, mat, chopped and milled
(Irwindale, Cal.)	Kaiser Glass Fiber Corporation	AeroROVE (chopped strand, roving and woven roving)

Table 4.5—*continued*

Country	Producer	Trade name
(Waterville, Ohio)	Johns–Manville Fiber Glass Inc.	Garan, Garanmat, Vitro-Flex, Vitro-Strand, Vitron (yarn, roving, woven roving, mat, chopped and milled)
(Pompano Beach, Fla.)	Lundy Electronics & Systems Inc.	Metallized glass roving
(Farmingdale, NY)	The Oliver Glass Fiber Corporation	Yarn
(Huntingdon, Pa., Ashton, R. I., Aiken & Anderson, S.C. and Jackson, Tenn.)	Owens–Corning Fiberglas Corporation	Fiberglas, Activa, Beta (yarn, roving, woven roving, mat, chopped, milled and staple yarn and sliver)
(Lexington & Shelby, NC)	PPG Industries, Inc. Fiber Glass Division	Feneshield, Hybon, Hycor, Pittsburgh, PPG, Triant (yarn, roving, mat, chopped and milled)
(Bremen)	Reichhold Chemicals Inc., Glass Fiber Division	Modiglass
(Statesville)	Uniglass Industries, A Division of United Merchants & Manufacturers, Inc.	Uniglass

4.2.4 Forms and Properties of Glass Fibres for Use in Reinforced Thermoplastics

Glass fibres are supplied for use as reinforcement in thermoplastics in two physical forms, i.e. continuous roving and chopped strand.* Rovings

* The appropriate British Standard definitions (BS 3691 : 1969) are: roving—a plurality of parallel strands; strand—a plurality of filaments bonded with a size.

are bundles of strands, resembling untwisted rope, formed in a secondary operation, after the production of the strands, and wound to form a cylindrical package. As implied by their name, chopped strands are glass filament strands cut into pre-determined lengths, normally 3, 6 and 12 mm ($\frac{1}{8}$, $\frac{1}{4}$ and $\frac{1}{2}$ in). Such 'glass variables' as filament diameter, number of filaments in a strand, length of fibre in the composite, type and amount of size and/or coupling agent, all affect the properties of glass-reinforced thermoplastic materials. Some of the effects were studied by Maaghul.[7] Results of an examination of these factors in conjunction with compounding conditions, moulding conditions and glass loading in a number of thermoplastics have been reported by Richards and Sims.[8] Useful information on the nature and effects of glass-fibre reinforcement in thermoplastics has been published by Crabtree and Pickthall.[9]

It should be noted that, depending on the compounding process used (*cf.* Chapter 5) and the fabrication method (*cf.* Chapter 7), the glass fibre reinforcement finally present in a filled thermoplastic moulding material, or in the moulding itself, may consist either of long fibres or short ones or a mixture of the two. Final length and length distribution in the moulding is largely independent of the initial form of the glass, i.e. whether continuous roving or chopped strand, but is essentially a function of the type and severity of treatment during processing (compounding and moulding). Glass fibres recovered (by burning off the resin) from melt-compounded polymer granules are shown in Fig. 4.2.

In practice virtually all the glass fibre reinforcement in commercial reinforced thermoplastics is fibrous E glass. Apart from some grades of asbestos it is the cheapest high-performance fibrous reinforcement for thermoplastics.

4.2.5 Coupling Agents

In the course of development of thermosetting, glass-reinforced plastics compositions, it was realised that the bonding between the polymer matrix and the glass reinforcement could be improved by treating the glass fibre surface with a very thin layer of a reagent which would form an adhesion bridge between the reinforcement and the polymer. For this reason such agents are called coupling agents. Coupling agents are used, and fulfil the same function, in glass-reinforced thermoplastics. The way in which improved adhesion between the fibrous reinforcement and the matrix enhances the strength properties of the reinforced thermoplastic material has been considered in Chapter 1, Section 1.4. An additional advantage of the improved adhesion which results from the use of appropriate coupling agents is a possible reduction in the water absorption of the reinforced compositions because inward penetration of water along the glass/polymer interface may be reduced.

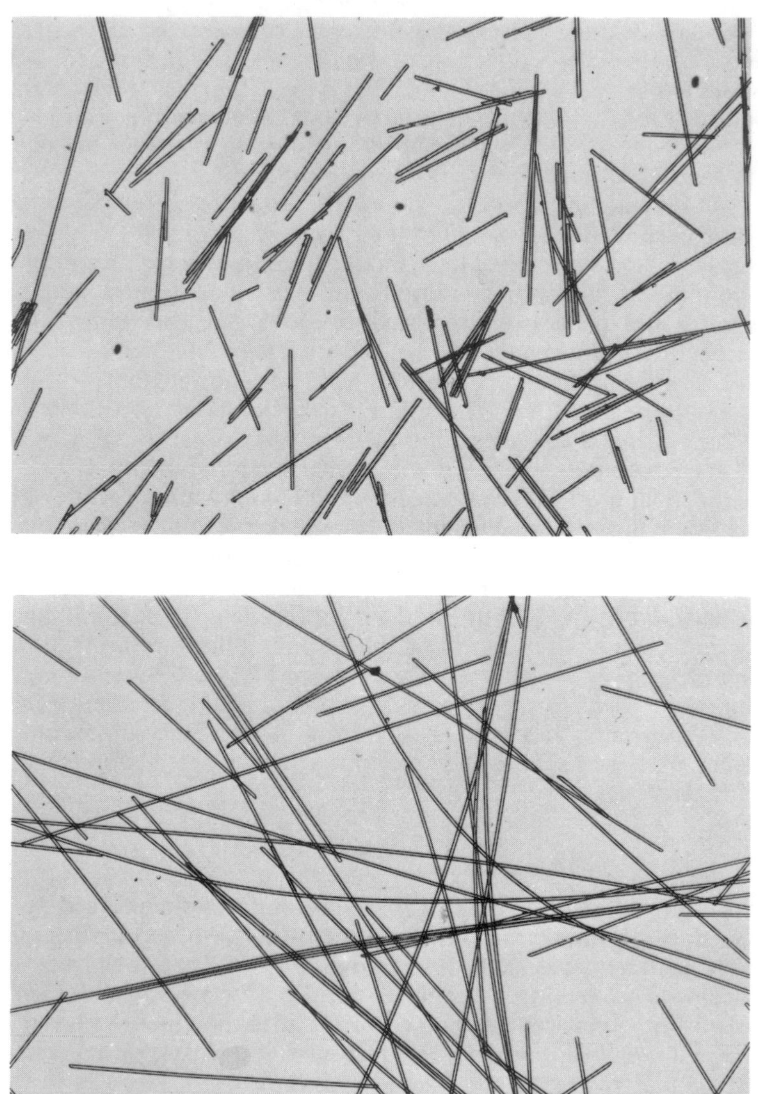

FIG. 4.2 Glass fibres recovered from melt-compounded polymer granules (compounding conditions appropriate to the polymer filled). (a) Nylon 6.6 (30% glass) (×31). (b) Polypropylene (40% glass) (×31).

It was realised quite early in the development of reinforced thermoplastics that the type of coupling agent originally found to be effective in glass-reinforced thermosets, especially certain silane compounds, was also by and large effective in reinforced thermoplastics.[10] Table 4.6 shows some of the commercial silane coupling agents most of which are suitable for application to glass reinforcement for thermoplastic polymers.

Chromium compounds, e.g. Volan A—a methacrylate chromic chloride complex, were among the earliest coupling agents used in glass-reinforced thermoset materials. However, for reinforced thermoplastics, silane coupling agents are the type most commonly employed.

The way in which silane coupling agents act, and in particular the bonding mechanism to the glass fibre surface on one side and the polymer matrix on the other, have been the subject of considerable interest and are still under study. There is as yet no complete and fully established explanation. Available results suggest that the mechanism is complex and that it may be different in different systems. In discussing the performance of XZ-8-5069 (Dow Corning)—a silane derivative of an unsaturated amine hydrochloride—Plueddemann[10] comments that its high effectivity as a coupling agent for reinforcing fillers in most thermoplastic (as well as thermosetting) resins, albeit not fully understood, probably depends on the effects of several structural features of the molecule combining and mutually reinforcing.

Some investigators believe that actual covalent bonding of the silane to both the polymer matrix and the reinforcing filler surface is an important factor in the coupling effect, and possible reaction mechanisms have been suggested.[9,11,12] 'Solution compatibility' between the silane and some glassy polymers, manifested in good wetting of the silane-treated glass surface by the polymer, and in similarity of the solubility parameters of the silane and polymer, has been put forward as a factor.[13,14] Grafting and cross-linking have also been suggested as mechanisms of coupling of polypropylene and polystyrene to glass via organofunctional silanes.[10]

Significant improvements in the strength of glass-fibre/polystyrene and glass-fibre/SAN composites were obtained by Hartlein[15] using a two-component coupling system:* a silane coupling agent applied to the glass-fibre surface and overcoated with a tough thermosetting polymer applied from solution. Two main reasons may account for the success of this approach: intimate interfacial contact between the polymer coating and the silane layer on the glass, and the protection afforded by the coating to this layer, which must reduce disruption of the latter in processing and hence loss of the coupling effect.

From the practical standpoint, an effective coupling agent must,

* Similar ideas have also been put forward by Edwards—see Appendix 1, British Patent 1 095 700.

TABLE 4.6
Some Silane Coupling Agents

Name	Composition	Source and designation of commercial product	Polymers for which recommended by suppliers
Vinyltrimethoxysilane	$CH_2{=}CHSi(OCH_3)_3$		
Vinyltriethoxysilane	$CH_2{=}CHSi(OC_2H_5)_3$	A-151 (Union Carbide)	Thermosetting polyester
Vinyltrichlorosilane	$CH_2{=}CHSiCl_3$	A-150 (Union Carbide)	Thermosetting polyester
Amyltrimethoxysilane	$C_5H_{11}Si(OCH_3)_3$		
Phenyltriethoxysilane	$C_6H_5Si(OC_2H_5)_3$	Z-6071 (Dow Corning)	
β-cyclohexylethyltrimethoxysilane	$C_6H_{11}CHCH_2CH_2Si(OCH_3)_3$ (ring: H_2C, C—C with H_2 H_2, CH)		
γ-methacryloxypropyltrimethoxysilane	$CH_2{=}C(CH_3){-}C({=}O){-}O(CH_2)_3Si(OCH_3)_3$	A-174 (Union Carbide)	Polystyrene, polyethylene, ABS, polypropylene (and thermosetting polyesters)
γ-iodopropyltrimethoxysilane	$I(CH_2)_3Si(OCH_3)_3$	XZ-8-5021 (Dow Corning)	
γ-chloropropyltri-		XZ-8-0999	

γ-chloroisobutyltri-ethoxysilane	$ClCH_2CH(CH_3)CH_2Si(OC_2H_5)_3$		
γ-glycidoxypropyltri-methoxysilane	$\underbrace{CH_2—CH}_{O}CH_2O(CH_2)_3Si(OCH_3)_3$	Z-6040 (Dow Corning) A-187 (Union Carbide)	ABS, SAN (and some thermosets)
γ-aminopropyltri-ethoxysilane	$H_2N(CH_2)_3Si(OC_2H_5)_3$	A-1100 (Union Carbide)	PVC, polycarbonate, nylon, polypropylene, polymethylmethacrylate
N-β-aminoethyl-γ-amino-propyltrimethoxysilane	$H_2N(CH_2)_2NH(CH_2)_3Si(OCH_3)_3$	Z-6020 (Dow Corning) A-1120 (Union Carbide)	Epoxy, phenolic
N-*bis*-(β-hydroxyethyl)-γ-aminopropyltri-ethoxysilane	$(HOCH_2CH_2)N(CH_2)_3Si(OC_2H_5)_3$	A-1111 (Union Carbide)	PVC, nylon, polysulphone (and epoxy)
β-(3,4-epoxycyclohexyl)-ethyltrimethoxysilane	3,4-epoxycyclohexyl ring (O bridging CH–CH; ring CH_2 groups) $—CH(CH_2)_2Si(OCH_3)_3$	A-186 (Union Carbide)	Polystyrene, ABS, SAN

according to Hartlein,[15] 'be of high modulus, be non-melting, and chemically adhere to the glass fibre surface . . . must not be removed during moulding and must present a surface chemically similar to the matrix resin or at least one to which the matrix resin adheres tenaciously'.

As has been mentioned, the coupling agent is normally applied to the freshly formed glass fibres† as a component of the size (often called 'forming size'). Commercially used sizes are water based and typically contain, apart from the coupling agent:

(a) a film-forming polymer: this may be e.g. polyvinyl acetate, starch or polyester, emulsified in the aqueous phase: it gives protection to the glass filaments and cohesion to the strand;
(b) plasticiser(s) or modifier(s) for the polymer (e.g. dibutyl phthalate if the polymer is polyvinyl acetate);
(c) a lubricant (e.g. a silicone) to facilitate drawing and reduce friction against guide points in winding;
(d) antistatic agents.

The water content of the size may typically be about 94%.

4.3 GLASS BEADS (MICROSPHERES)

4.3.1 Types, Nature and Size

Two kinds of glass beads are available for use as fillers in plastics: solid beads, often called Ballotini, and hollow microspheres (glass microballoons). Only the former function as a true reinforcing filler and are used as such in commercial reinforced thermoplastics.

4.3.1.1 Hollow Microspheres

The hollow microspheres are individually colourless, but in bulk they appear as a buff-coloured, free-flowing, sand-like powder. This light-weight filler is available in two density and wall thickness ranges.[16,17]

The interior of the microspheres contains a gas. In the Armoform spheres this has been identified by Raask[18] as a mixture of mainly nitrogen and carbon dioxide.

Hollow microspheres are not a reinforcing filler in the true sense. Their main uses so far have been in syntactic foams based on thermosetting resins, and as a filler to reduce simultaneously the density and the dielectric constant of epoxy resins used as encapsulants in electrical applications.[16,17]

4.3.1.2 Solid Microspheres (Ballotini)

Ballotini appear transparent under the microscope but white in bulk (see Fig. 4.3).

† Addition to the polymer has also been proposed—see, e.g., Appendix 1, British Patent 1 131 533.

	Thick-walled ('Armoform' spheres—Microshells Ltd., UK)	Thin-walled ('Eccospheres'—Emerson and Cuming Inc., USA and UK)
Particle size (μm)	20–200	100 (average)
Density (g cm^{-3})		
wall material	About 2·6	About 2·6
single sphere (wt/vol)	0·4–0·6	About 0·21
bulk	0·25–0·40	About 0·13
Wall thickness (as percentage of diameter)	About 10	About 1·5

They can be produced in various particle sizes. For example six standard size ranges (as well as special ones) are offered by Plastichem Ltd. in the UK.

The 4–44 μm size range (Ballotini 3000) is the grade most commonly used in plastics materials, both thermoplastic and thermoset. The particle size distribution for this grade is shown in Fig. 4.4.[19]

The material of solid glass beads is type A glass, not type E as used in the glass-fibre reinforcement for thermoplastics. This is because E glass melt does not readily form spherical droplets in the modern manufacturing process (*cf.* Section 4.3.2). Special methods, which would be necessary to

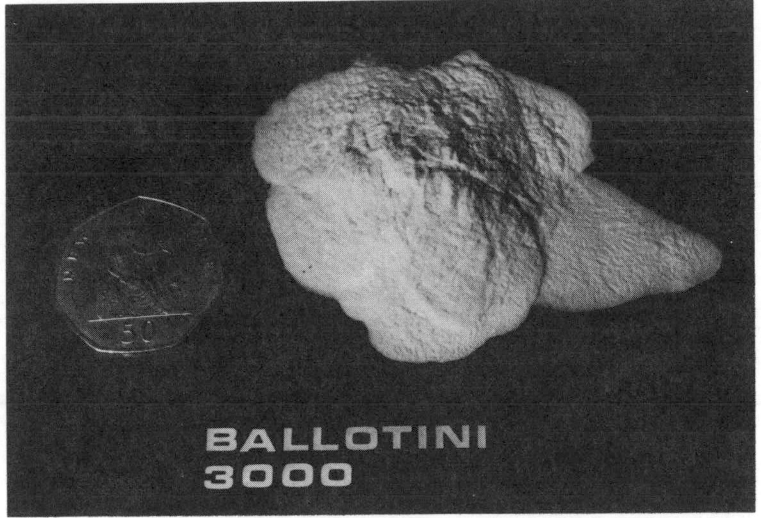

FIG 4.3 Ballotini: bulk appearance. (Courtesy of Plastichem Ltd., England.)

Code	Particle size range (80% within range stated) (μm)	(in)	US mesh nos. (ASTM E-11)
1221	177–840	0·0070–0·0331	80–20
1721	177–354	0·0070–0·0139	80–45
1922	149–250	0·0059–0·0098	100–60
2024	105–210	0·0041–0·0083	140–70
2429	53–105	0·0021–0·0041	270–140
3000	4–44	<0·0017	>325

achieve satisfactory 'beading' of E glass would make production, and hence the product, more expensive. This is not commercially justifiable since the functional properties of E glass beads would not be significantly superior to those of the standard glass beads.

4.3.2 Manufacture and Properties of Glass Spheres

The glass (in any form including broken pieces) from a glass works is crushed in a hammer or ball mill, sieved and fed to a special furnace in which it is carried upwards, whilst melting, in a stream of hot gases produced by an annular flame at the bottom of the furnace. Molten spheres are formed en route and are removed when they reach the top. Grading starts in the course of this operation in that any over-size spheres fall back to the bottom. Coupling agents are normally applied directly after

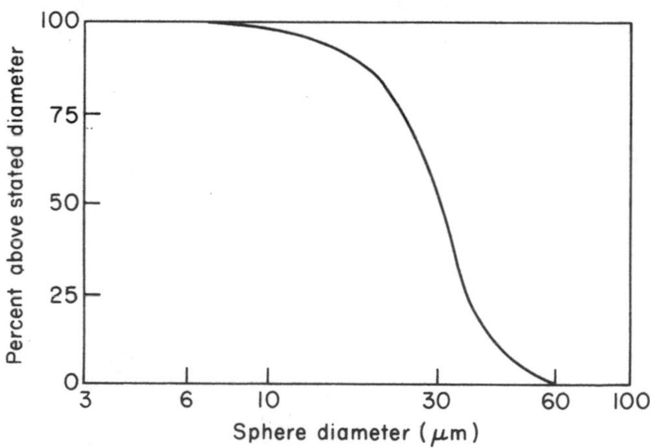

FIG. 4.4 Particle size distribution of Ballotini 3000 grade. (Strauch's data: reprinted, with permission, from *SPE Journal*, **25**, 9, September 1969.)

formation, to the clean, fresh surface. Several of those shown in Table 4.6 are suitable. Many coupling agents recommended by Ballotini suppliers as suitable for their products are proprietary versions of these compounds. As pointed out by Hopkins,[20] the physical properties of a composite reinforced with glass spheres are improved by about 10% if the spheres have been treated with a suitable coupling agent.

The firm interfacial adhesion between a surface-treated glass sphere and a thermoplastic polymer matrix (modified polyphenylene oxide) is illustrated by the electron-scan micrograph of Fig. 4.5.

After formation and any surface treatment the spheres are tested for degree of roundness, and the particle size is determined. The former test is also a grading operation, carried out on a vertically vibrating inclined plane which permits only those spheres with a satisfactory degree of roundness to roll to the bottom. Final sieving completes the grading process. Particle size determination is normally also by sieving and/or microscope inspection.

The properties of the glass used as starting material are tested to ensure that the quality is suitable and constant. Tests are also carried out on the final product. Some typical property values are shown in Table 4.7.

The main supplier of glass beads is the Potters-Ballotini group of

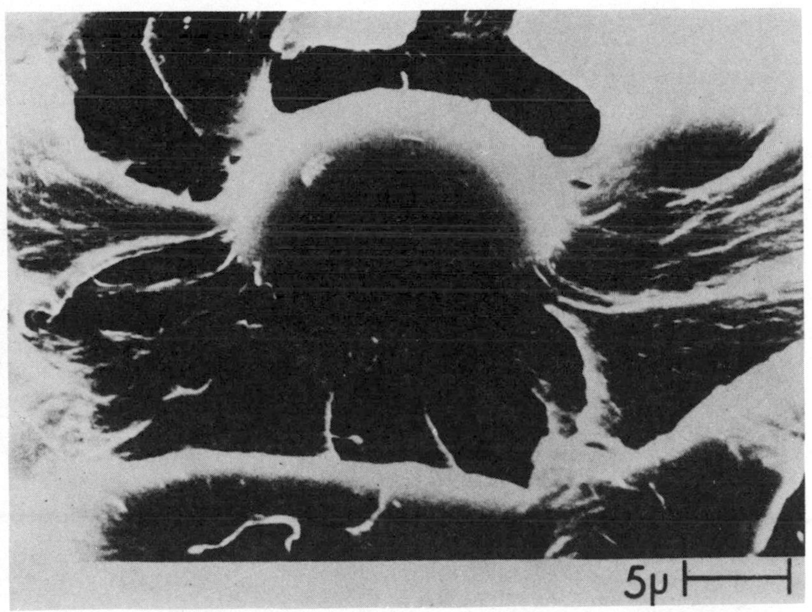

FIG. 4.5 Glass sphere 'coupled' to a PPO matrix (specimen under tension). (Courtesy of Plastichem Ltd., England.)

companies, which includes Potters Industries Inc. in the USA, Potters-Ballotini Ltd. in the UK and other manufacturing companies in Belgium, Germany, Canada and Mexico. Ballotini-reinforced nylon moulding compounds ('Well-sphere') and compounds containing both glass spheres and fibre ('Well-A-Meld') were produced, until recently, by Wellman Inc. in the USA and their associates abroad. Production has now ceased because of raw-material shortage.

TABLE 4.7
Properties of Glass Spheres
(Courtesy of Plastichem Ltd.)

Property	Value and units
Physical and general	
Specific gravity	2·46–2·49
Refractive index	1·51–1·52
Young's modulus	1×10^7 psi
Rigidity modulus	$4{\cdot}3 \times 10^6$ psi
Poisson's ratio	0·21
Coefficient of friction	0·18 to 0·24 (glass on glass)
Hardness	
DPH 50 g load	540 kg mm^{-2}
Knoop 100 g load	515 kg mm^{-2}
Thermal	
Strain point	505°C
Annealing point	548°C
Softening point	730°C
Expansion coefficient (0–300°C)	85×10^{-7} per °C
Mean specific heat	0·18 at 20°C 0·28 at 1 000°C cal g^{-1} $°C^{-1}$
True specific heat	0·18 at 20°C 0·32 at 1 000°C cal g^{-1} $°C^{-1}$
Thermal conductivity	0·002 at 0°C 0·0036 at 500°C cal s^{-1} cm^{-1} $°C^{-1}$
Thermal diffusivity at room temperature	0·005 cm^2 s^{-1}
Electrical	
Dielectric constant (1 MHz at 20°C)	7·0–7·6
Power factor (1 MHz at 20°C)	0·004–0·011
Volume resistivity at 250°C	$10^{6\cdot5}$–$10^{7\cdot0}$ Ω
Dielectric strength	4 500 kV cm^{-1}
Chemical durability	
Powder tests	
In water, 4 h at 90°C	0·05% Na_2O extracted
In N/50 H_2SO_4, 4 h at 90°C	0·03% Na_2O extracted

4.3.3 Effect in Thermoplastics

The advantages claimed for solid glass spheres as a filler for thermoplastics fall under three headings: processing, property improvement, and economy.

The main processing advantage is that—especially at high filler loadings—the viscosity of the melt is increased to a much lesser extent by glass spheres than it is by irregular particles (e.g. clay, powdered quartz) or glass fibres.[20,21] This makes production of complex or thin-walled mouldings easier with sphere-filled than with glass-fibre-filled thermoplastics.[20,22] Nylon 6.6 moulding compounds filled with glass spheres or glass-sphere/glass-fibre mixture (Well-Sphere and Well-A-Meld) have been found to require considerably lower melt temperatures than a comparable glass-fibre-reinforced compound in the injection moulding of an intricately shaped gear.[22]

The principal effects on properties in comparison with the base polymer include reduction of shrinkage, improvements in abrasion resistance (up to 750% increase has been claimed in ABS cups for an automatic poultry watering system[20]), compressive strength, hardness, tensile strength, modulus and creep. In general the improvements in strength, modulus and creep are less than those produced by glass fibre at a comparable level of loading; however, because of the regular shape of the spherical filler, the improvement in compressive strength is correspondingly greater, and the reinforcing effect is isotropic since filler orientation cannot take place. Good surface finish can also be obtained in comparison with glass-fibre-filled materials.

The cost of glass spheres is lower than that of glass fibres or the more expensive polymers. For 40% sphere-filled nylon 6.6, a recent (November 1973) calculation based on average USA prices indicated a cost saving of approximately 16% over the corresponding glass-fibre-filled compound.[23] On the basis of current (mid-1974) UK prices this saving would be about 21%, and the saving in comparison with unfilled polymer about 14%. This is illustrated by the example below (cost figures rounded off):

Polymer at £1000 per tonne — Density* 1·15 g cm^{-3}
Fillers
  Glass spheres at £200 per tonne — Density* 2·47 g cm^{-3}
  Glass fibres at £600 per tonne — Density* 2·55 g cm^{-3}
Filler loading 40% by weight (≡66 phr†)

The polymer is sold and compounded on a weight basis, but it, and any filled compound prepared from it, is *used* on a volume basis. If a tonne of

* Numerically equal to specific gravity.
† Parts by weight of filler per hundred parts of resin.

unfilled polymer produces say 4000 mouldings, each of which is a fixed *volume* item, then the amount of reinforced compound equivalent for price comparison purposes will be the *volume* of compound to give the same number of mouldings, i.e. the volume of compound equal to the volume of a tonne of unfilled polymer (differences in shrinkage are neglected for the purpose of this illustration).

The volume fraction of polymer in the filled compound will be:

$$\frac{W_p}{\rho_p} \Bigg/ \left(\frac{W_f}{\rho_f} + \frac{W_p}{\rho_f}\right) = \frac{\rho_f W_p}{\rho_p W_f + \rho_f W_p}$$

where W_p, W_f = weights of polymer and filler respectively, ρ_p, ρ_f = densities of polymer and filler respectively.

Thus in this example the volume fraction of polymer in 40 wt % sphere-filled compound is:

$$\frac{2{\cdot}47 \times 100}{1{\cdot}15 \times 66 + 2{\cdot}47 \times 100} = \frac{247}{76 + 247}$$

$$= 0{\cdot}765$$

Hence the volume of compound equal to that of one tonne of polymer will contain:

0·765 tonne of polymer

costing 0·765 × £1000 = £765

The filler : polymer ratio by weight is 66 : 100, hence the compound will contain:

$$\frac{66}{100} \times 0{\cdot}765 \text{ tonne of filler}$$

costing 0·66 × 0·765 × £200 = £101

The cost of the compound will be:

£765 + £101 = £866 per 'tonne equivalent'

giving a saving of:

£1000 − £866 = £134,

i.e. about 13·5% saving on a tonne of polymer.

Analogous calculation for 40% fibre-filled compound gives the cost of compound:

£1075 per 'tonne equivalent',

i.e. a saving for spheres against fibres of:

£1075 − £866 = £209 (≡ 21% approximately)

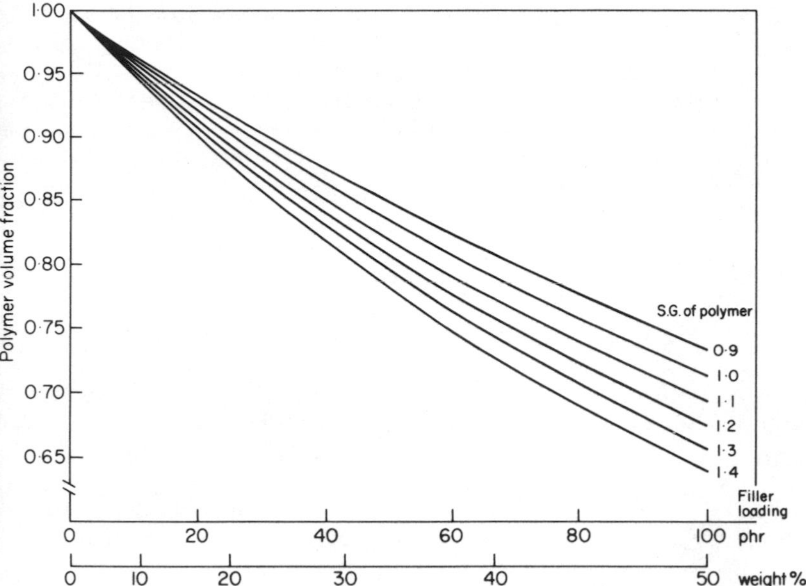

FIG. 4.6 Volume fraction of polymer *vs* glass loading (average specific gravity of 2·5 assumed for both spheres and fibres).

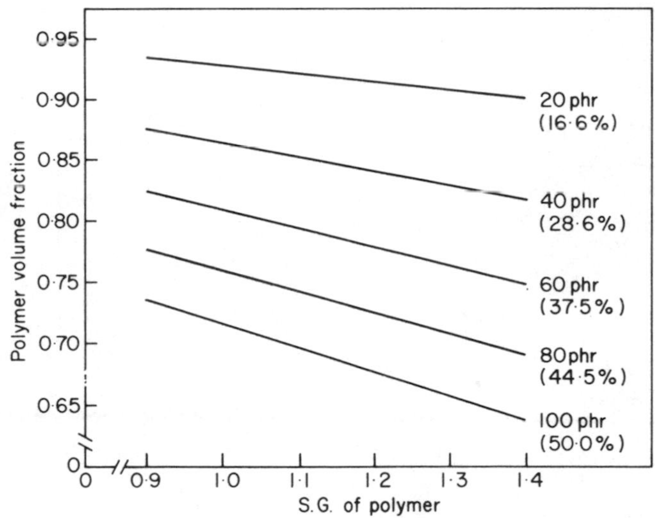

FIG. 4.7 Volume fraction of polymer *vs* specific gravity at constant levels of glass loading (average specific gravity of 2·5 assumed for both spheres and fibres).

The graphs in Figs. 4.6 and 4.7 show the relationships between the volume fraction and specific gravity of the polymer and the reinforcement loading in compounds reinforced with glass fibres or spheres. The same specific gravity (2.5) has been assumed for both fillers, but thc approximation is sufficiently close for the graphs to be relevant (actual specific gravity values: E-glass fibres 2·54–2·56; A-glass spheres 2·46–2·49 —*cf.* Tables 4.4 and 4.7).

4.4 ASBESTOS FIBRES

4.4.1 Nature, Origin and Types of Asbestos

Asbestos is the general name given to a group of fibrous silicates of natural, mineral origin. The general classification and chemical composition of asbestos fibres are schematically shown in Fig. 4.8.

As has been mentioned in Section 4.1, asbestos fibres are crystalline, whereas glass fibres—the man-made silicate fibres widely used as reinforcement in thermoplastics—are amorphous. It is for this reason that, whilst the two types of fibre are roughly comparable in tensile strength, asbestos fibres are much stiffer (*cf.* Table 4.1). The general differences in fibre length and fineness between glass and asbestos fibres are also the result of their respective origins. In asbestos deposits the fibres are of finite length; by longitudinal splitting, which involves cleavage along crystal planes, of the original rock fragments or coarse fibre aggregates, they can be obtained as individual fibres with diameters of 0·1 μm (amosite, crocidolite and

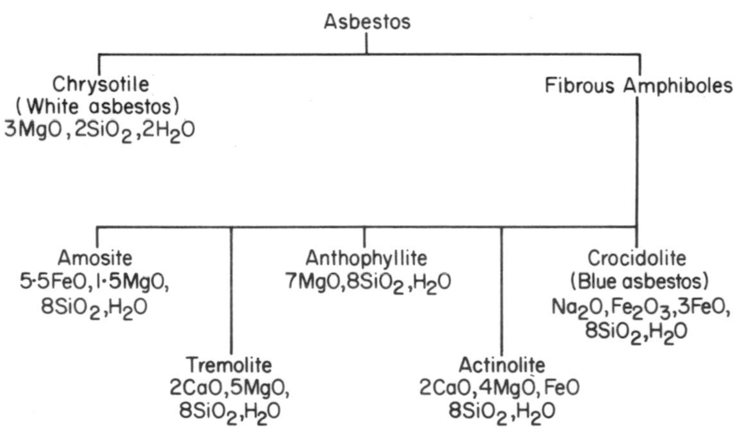

FIG. 4.8 Varieties of asbestos—after Hodgson.[3] (Courtesy of the Royal Institute of Chemistry.)

FIG. 4.9 Crocidolite fibres partly separated from the parent rock. (Courtesy of Cape Asbestos Fibres Ltd.)

anthophyllite) or even finer, down to 0·03–0·01 μm (chrysotile). The fineness of the fibres makes for high aspect ratios and hence good reinforcing effects (*cf.* Chapter 1, Section 1.4).

Figure 4.9 illustrates the position of asbestos fibres in the native rock and the increasing fineness with progressive separation.

The texture of asbestos is an important property, reflecting the fundamental properties of the fibre type, and influencing the processing and applications. Textures are normally designated as harsh, semi-harsh and soft. They vary with the type of asbestos. The types containing a high proportion of magnesium (chrysotile, anthophyllite, tremolite) tend to be of soft texture (semi-harsh in some varieties of chrysotile) with white, silky fibres. The texture of crocidolite (blue asbestos) is semi-harsh and the fibres resilient: amosite is a harsh asbestos with resilient spiky fibres.[3,24] Both these types of asbestos contain a high proportion of iron.

The differences in texture, as well as in the general chemical and surface properties, are functions not only of the chemical composition but also of crystal structure: chrysotile differs greatly from the amphibole group in this latter respect, as shown schematically in Figs. 4.10 and 4.11.

Some of the properties of representative asbestos types are shown in Tables 4.8 and 4.9.

TABLE 4.8
Physical and Physico-Chemical Properties of the Asbestos Minerals
(Courtesy of Cape Asbestos Fibres Ltd.)

Property	Amosite	Crocidolite	Chrysotile
Colour	Ash grey	Blue	White
Natural fibre length (mm)	3–70	3–70	1–40
Fibre diameter (μm)	0·1–1	0·1–1	0·01–1
Crystal system	Monoclinic	Monoclinic	Monoclinic and Orthorhombic
Occurrence	Banded ironstones	Banded ironstones	Serpentines
Optical properties	Biaxial positive extinction parallel	Biaxial positive or negative extinction parallel	Biaxial positive extinction parallel
Refractive index	1·657–1·698 non-pleochroic	1·685–1·698 pleochroic	1·50–1·55 non-pleochroic
Specific gravity	3·2–3·45	3·37	2·55
Hardness (Mohs' scale)	5·5–6·0	4·0	2·5–4·0
Specific heat (kcal g^{-1} °C^{-1}	0·193	0·201	0·266
Tensile strength (kg cm^{-2})	25,000	35,000	31,000
Young's modulus (kg cm^{-2})	$1{\cdot}65 \times 10^6$	$1{\cdot}9 \times 10^6$	$1{\cdot}65 \times 10^6$
Heat resistance:			
% residual strength at 200°C	100	100	100
% residual strength at 400°C	37	38	100
% residual strength at 600°C	7	18	16
Fusion point °C	1 100	1 000	1 500
Surface charge	Negative	Negative	Positive

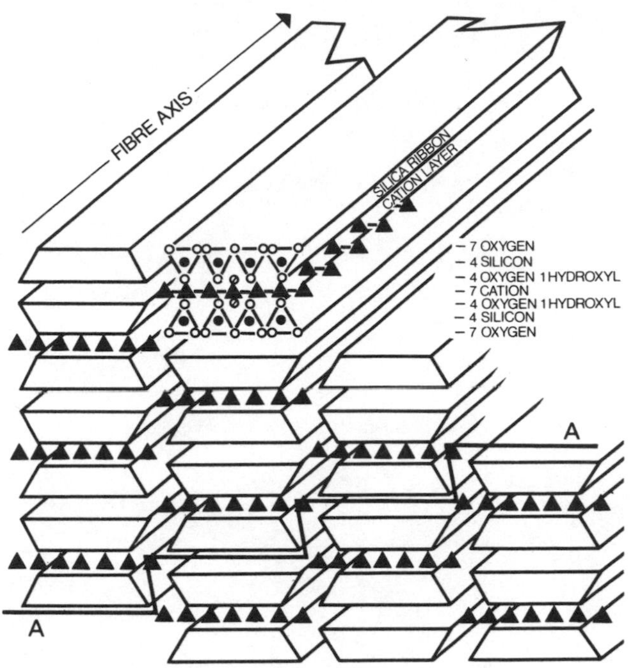

FIG. 4.10 Schematic diagram of the crystal structure of an amphibole fibre, indicating the unit cell $X_7Si_8O_{22}(OH)_2$. The line A–A represents the edge of the preferred cleavage plane along which the fibres will split to form even smaller fibres. (Courtesy of Cape Asbestos Fibres Ltd.)

TABLE 4.9

Chemical Compositions of the Asbestos Minerals: Typical Data

(Courtesy of Cape Asbestos Fibres Ltd.)

	Amosite	Crocidolite	Chrysotile
SiO_2, silica	49·7	50·9	38·9
FeO, ferrous oxide	39·7	20·5	2·0
Fe_2O_3, ferric oxide	trace	17·45	1·6
Al_2O_3, alumina	0·4	—	3·1
CaO, calcium oxide	1·0	1·5	0·9
MgO, magnesium oxide	6·4	1·0	40·0
MnO, manganese oxide	0·2	0·05	0·1
Na_2O, sodium oxide	0·1	6·2	0·1
K_2O, potassium oxide	0·6	0·2	0·2
H_2O, water	1·8	2·0	12·6
CO_2, carbon dioxide	0·1	0·2	0·5
	100·0	100·0	100·0

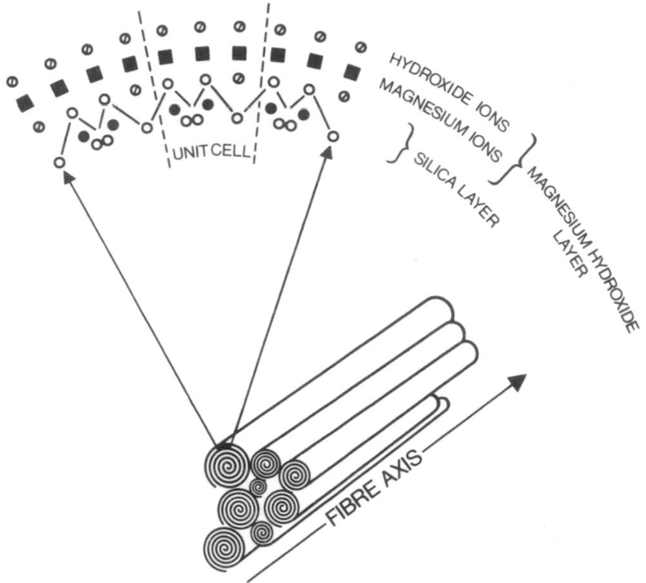

FIG. 4.11 Schematic diagram of the structure of a chrysotile fibre formed of several scrolls of individual crystallites. Each scroll is formed from a closely connected double layer having magnesium hydroxide units on its external face and silica units on its inner face. The details of a small section of the scroll show the structure of the double layer and of the unit cell based on $Mg_3(Si_2O_5)(OH)_4$. (Courtesy of Cape Asbestos Fibres Ltd.).

4.4.2 Production of Asbestos Fibres

Asbestos ore is mined: chrysotile mainly by open-cast methods and the amphibole group normally by deep-mining. Ore delivered to the mill is crushed, washed and where possible barren rock pieces are separated out. After further crushing processes fibre and rock fragments are separated and graded by means of a series of shaking screens with air lifts. After a final degritting process the cleaned and graded fibre is pressure-packed into impermeable bags.[24] It is common, for economic reasons, for these operations at the mine to be aimed at producing a semi-crude fibre form, which is finally processed and graded after shipment. The processing is essentially further fiberisation, to separate out the fibres more completely, with, as far as possible, minimum reduction in length. Quality tests are applied to determine and control such important characteristics as fibre length (usually by screening, wet or dry), fibre diameter (usually through determination of the permeability, and hence fibre surface area, in a standard plug), and the degree of fiberisation.[24]

The main British and North American suppliers of asbestos fibre to the plastics industry are listed below:

UK
Cape Asbestos Fibres Ltd., Uxbridge, Middlesex, England.
Turner Brothers Asbestos Co. Ltd., Rochdale, Lancashire, England.
North America
Asbestos Corp., Thetford Mines, Que., Canada.
Cal-Fiber Co., Los Angeles, California, USA.
North American Asbestos Corp., Chicago, Illinois, USA.
Raybestos Manhattan Inc., Bridgeport, Connecticut, USA.
Johns Manville Corp., Denver, Colorado, USA.

4.4.3 Asbestos Fibres Used as Reinforcement in Thermoplastics

There are two major applications of asbestos fibres in a thermoplastic, which do not involve injection moulding. Both are in PVC. Fairly high loadings of asbestos fibre are incorporated in certain flooring compositions (calendered sheet): the fibre serves mainly as a filler (extender), although some reinforcement also results. Asbestos (chrysotile)-reinforced PVC homopolymer sheet is also produced, by calendering, from a solvent-containing dough. The process produces planar, as well as some directional, orientation of the fibres, and true reinforcement is obtained.[25]

Asbestos fibres are used as reinforcement in thermoplastic moulding compounds, mainly in polypropylene, and—to a lesser extent—in HD polyethylene, polystyrene and nylon. In these materials the asbestos fibres improve the stiffness and the strength of the composites. They also upgrade the heat distortion temperature in many cases and reduce shrinkage.

In polypropylene, asbestos has a deleterious effect on the heat stability of the polymer. Anthophyllite has been early and widely used because it is the least active in this way. The degradative effects of chrysotile are now to some extent controllable and it is being used as reinforcement in polypropylene moulding compounds. Even with anthophyllite it is usual to employ stabilisers which inhibit the thermal and oxidative degradation of the polymer. Various inhibitors are used, some based on phenol derivatives in conjunction with epoxy resins.[26] Pre-extraction of the fibre with acids has also been claimed to improve stability and interfacial adhesion in the composite.[27] By and large, the amphibole fibres are easier to compound with the polymer than chrysotile fibres, due mainly to differences in the chemical nature of the surface.

In general amosite gives better reinforcement than anthophyllite in polyolefins, but its adverse effect on polymer stability is marked, especially in polypropylene. Recent work at Cape Asbestos Fibres Ltd. is reported to have provided practicable means of overcoming this difficulty.[28] Amosite reinforcement in polypropylene matrix is shown in Fig. 4.12, and the centre of an area discoloured through heat degradation (surface of amosite-reinforced polypropylene) in Fig. 4.13.

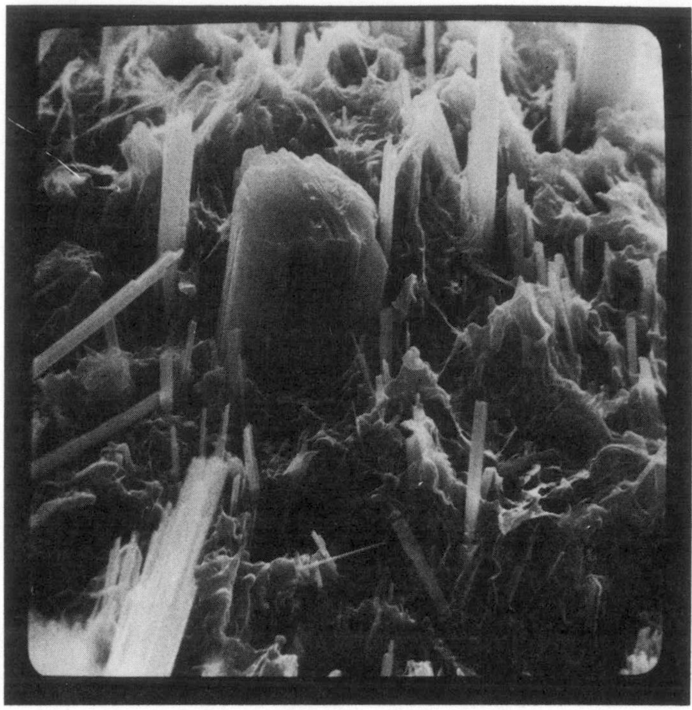

FIG. 4.12 35% amosite fibre reinforcement in polypropylene; electron scan: ×1680. (Courtesy of Cape Asbestos Fibres Ltd.)

Crocidolite fibres are potentially capable of giving the best reinforcement, but compounding these fibres into polymers requires special health-protective precautions.

As would be expected, longer fibres give better reinforcement but they also cause greater melt viscosity increase, for the same level of loading. These points are illustrated by the data of Table 4.10.

In general, shorter-fibre grades within a given type of asbestos are considered more suitable for compounding with thermoplastics. The aspect ratio remains generally good, even after partial breakdown in mixing, as illustrated by Fig. 4.14.

Asbestos dust (fibre fragments) is a health hazard: serious lung conditions may be caused by inhalation of excessive amounts over a long period of time.[30] However, safe working is possible.* The UK Asbestos

* Two useful sources of information are: 'Asbestos—Safety and Control', published by The Asbestos Information Committee, 10 Wardour Street, London W1V 3HG; and 'Asbestos and You', published by the Employment Medical Advisory Service, Department of Employment, Baynards House, Chepstow Place, London W.2.

TABLE 4.10

Effect of Asbestos Fibre Length on Degree of Reinforcement and Melt Flow in Polypropylene
(Data of Harrison and Sheppard.[29] Courtesy of The Plastics Institute.)

Property	Test method and units	Polypropylene with type and % of reinforcement shown		
		None	40% short-fibre asbestos	40% long-fibre asbestos
Flexural modulus	ASTM D-790 kgf cm^{-2}	18 000	49 000	70 000
Tensile strength	ASTM D-638(B) kgf cm^{-2}	350	350	600
Notched impact strength at 23°C	BS 2782 306A kgf cm cm^{-1}	8	4	6
Melt flow index (230°C/2·16 kg)	ASTM D-1238 —	3·0	1·0	0·1
Coefficient of linear thermal expansion	— (per °C)	11×10^{-5}	4×10^{-5}	2×10^{-5}

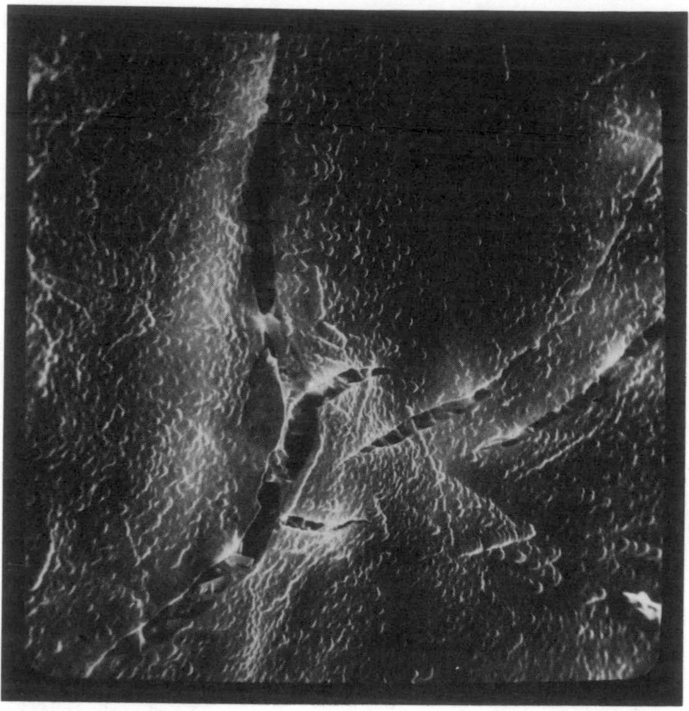

FIG. 4.13 Centre of heat-degradation spot on the surface of polypropylene/asbestos (40% amosite) moulding; electron scan: ×1600. Heat-treatment conditions leading to degradation: 376 h at 150°C. (Courtesy of Cape Asbestos Fibres Ltd.)

Regulations 1969 accept the principle of a threshold limit value (TLV) for asbestos dust concentration below which no special precautions need be taken. The values are 0·2 fibres per ml of air for crocidolite and 2 fibres per ml for other types of asbestos. Between the levels of 2 and 12 fibres per ml exposure to the atmosphere may be permitted in certain cases at the discretion of the Factory Inspectorate.[31] A good handling system for compounding polypropylene with asbestos fibres may be based on the following lines.

Asbestos fibre is delivered in woven polypropylene bags. The bags are slit and passed through a shredder, to be fed into the mix, whilst the fibre is fed, via a screw-and-hopper feed, into a blender into which the polymer (pre-mixed with stabiliser) is also simultaneously metered. The conveying system and space above the blender are dust-exhausted. The mix from the blender is transferred to a compounding machine (e.g. internal mixer, Buss Ko-Kneader, etc.—*cf.* Chapter 5) and the compound produced is subsequently pelletised. Direct feeding of 'dry' blends of asbestos fibre and

FIG. 4.14 Amosite fibres recovered from a polypropylene/amosite moulding (by leaching out polymer with H_2SO_4/HNO_3 mixture); electron scan: ×1600. (Courtesy of Cape Asbestos Fibres Ltd.)

polymer into the injection-moulding machine (*cf.* Chapter 5, Section 5.2.3.2) is not practised to any significant extent in Britain.

4.5 OTHER FIBROUS REINFORCEMENTS

Glass and asbestos fibres are the two main fibrous reinforcements used in commercial thermoplastic materials. In addition two other fibres are used, on a much smaller scale. These are potassium titanate whiskers, produced by DuPont under the trade name Fybex, and carbon fibres.

4.5.1 Whiskers

4.5.1.1 NATURE, ORIGIN AND PROPERTIES

As has been mentioned in Section 4.1, whiskers are essentially single crystals in fibrous form. They can be prepared from various metals and from certain metal compounds.[4] Whilst they are normally short their diameters are small (of the order of 0·1 μm) so that their aspect ratios can be high : about 40 : 1 or even higher is not uncommon. Exceptionally long sapphire whiskers (reputedly several metres long) have been prepared by a US Air Force laboratory,[2] not, however, by methods suitable for industrial production. Whilst various methods of preparation have been described, whiskers are most often prepared by vapour deposition.[2,4] Silicon carbide, silicon nitride, aluminium oxide (sapphire), magnesium oxide, beryllium oxide, carbon and potassium titanate are the whisker materials which have received the most attention in recent years.[1,2,32–34] Only potassium titanate whiskers have so far achieved commercial significance as a reinforcing filler for thermoplastics (see Section 4.5.1.2); others find their uses in special applications, mainly aerospace and military. Some whisker-forming compounds, e.g. silicon carbide, have been deposited on filamentary substrates, e.g. quartz, tungsten. This 'whiskerising' produces filaments of high stiffness and strength.[34] Tungsten filament has also been used as a substrate for the deposition of boron, to produce so-called boron filaments. The deposition imparts to the composite filament the high strength and rigidity of boron, and a highly oriented coating free from imperfections can be obtained in the vapour-phase reaction used to generate the boron *in situ* (reduction of boron trichloride with hydrogen). The diameter of the original tungsten filament is increased considerably. The resulting material is comparable in strength with glass fibres but its stiffness is higher by a factor of 4–5. Because the filaments are not particularly fine, and because they are produced in continuous lengths, they are particularly suitable for directional reinforcement of composites. So far their use in plastics has been limited to the reinforcement of special thermoset compositions (usually epoxy resins) with boron-filament tapes or fabrics. The mechanical properties of some whisker materials are given in Table 4.1.

Because of the methods of their production, whiskers are expensive: reinforcement of thermoplastics with these fibres has also not, in general, produced the great improvements which might have been expected *a priori*. This has been variously attributed to fibre breakdown, with consequent reduction of the aspect ratio, in compounding and moulding[1,33] and to difficulties in dispersion, possibly combined with poor original uniformity of the whisker material.[35] Both causes probably contribute. The commercial potassium titanate whiskers (Fybex: DuPont) are a partial exception. The supply of Fybex was discontinued in 1974 but its use is discussed in this section because of its technical interest.

Fybex fibres are suitable for incorporation in thermoplastics, in which they substantially improve the stiffness (more than glass fibre: less than carbon fibre) and strength (less than glass fibre). Fybex also confers other advantages in the composite (see below) and is considerably cheaper than other whisker materials.

4.5.1.2 Potassium Titanate Whiskers (Fybex) as Reinforcement in Commercial Thermoplastics

Some of the properties of Fybex whiskers are shown in Table 4.1. Others, also relevant to their application as reinforcing filler in thermoplastics are:[36]

Diameter	0·1–0·16 μm
Aspect ratio	about 40 : 1
Surface area (BET N_2 method)	7–10 m^2 g^{-1}
Refractive index	2·35

Providing that in the course of compounding of Fybex fibres with polymer the melt is allowed to wet them out before substantial shear is applied (and that reasonable care is taken in moulding or extrusion[36,37]), fibre breakdown is not a serious factor. This is demonstrated by the comparatively small effect of repeated complete re-processing on mechanical properties of Fybex-reinforced nylon (Table 4.11).

Whilst the aspect ratio of the Fybex fibres is high enough for good reinforcement effect, the fibres are short. This produces reinforcement and shrinkage effects more nearly isotropic in comparison with those arising with longer fibre reinforcement (e.g. glass fibres).[35] The surface finish is also much better for the same reason. Fybex-filled ABS, polypropylene, modified PPO and polysulphone can be electroplated.[37,38]

As suggested by its refractive index, Fybex is reasonably effective as a pigment. Replacement of titanium dioxide pigment by Fybex in some glass-fibre-reinforced thermoplastics improves physical properties whilst providing the required opacity.[35,36]

TABLE 4.11

70/30 Nylon 6.6/Fybex: Effect of Re-moulding (100% regrind) on Physical Properties

(P. G. Linsen and R. F. Regester, Proceedings of the 27th Annual Conference RP/C Institute. SPI, 1972, Paper 11-D: reproduced with permission of the authors).

	Mould cycle			
	1	2	3	4
Tensile strength (psi × 10^3)	19·6	17·3	17·3	18·4
Tensile modulus (psi × 10^6)	1·4	1·4	1·4	1·3
Elongation (%)	2·2	1·6	1·6	2·1
Flexural strength (psi × 10^3)	31·8	29·7	30·0	29·0
Flexural modulus (psi × 10^6)	1·3	1·2	1·2	1·2
Notched Izod impact (ft lb in^{-1})	1·4	1·5	0·9	0·8
Unnotched Izod impact (ft lb in^{-1})	11·8	12·0	9·9	11·2

The inclusion of Fybex (about 15% by weight) in rigid PVC compounds for the production, by extrusion, of profiles for outdoor use, window frames and rain-water goods also combines pigmentation with considerable reinforcement and improvement in dimensional stability.[35]

A number of Fybex-reinforced polymers were available commercially, e.g. from LNP and its licensees, when Fybex was being produced.

4.5.2 Carbon Fibres

The carbon fibres used currently as reinforcement were first incorporated in thermosetting resins,[39,40] where they are still being employed, in the form of continuous filaments or fabrics, to provide very high directional strength and stiffness at low composite weight in special applications. These fibres, which combine high modulus and high strength with comparatively low density, are the carbon fibres discussed in this section. They may be styled 'high-performance carbon fibres' to distinguish them from the older, low-modulus, low-strength carbon fibres normally prepared from cellulosic fibre precursors (see Table 4.12).

The high-performance carbon fibres have now become a valuable reinforcement also in thermoplastics,[41] although their commercial utilisation in these materials is still not very extensive. This is due partly to the price, which remains comparatively high, partly to the fact that surface-treated, chopped fibre tow specially suitable for incorporation in thermoplastics has only recently appeared, and partly also to the well entrenched position of glass reinforcement in thermoplastics for engineering applications.

4.5.2.1 Nature, Preparation and Properties

The high-performance carbon fibres are compared with other main filamentary forms of carbon in Table 4.12. As can be seen from the table, the normally low-performance carbon fibres prepared from cellulosic precursors can be improved by hot stretching in the course of production. The mechanism of carbonisation of cellulose in the preparation of low-performance carbon fibres (including the role of molecular orientation in the precursor, and orientation increases on hot stretching) was extensively studied by Tang and Bacon[42] *inter alia.* Many other precursors have also been examined, including various forms of pitch, some of which have been successfully used commercially in the production of low-performance carbon fibres. An informative review of these developments, as well as methods of preparation of high-performance carbon fibres, their properties and structure, is contained in the monograph by Gill.[40] Other literature sources deal with specific precursors and processes.[43–46] High-performance, and in particular high-modulus, carbon fibres are today commonly produced by oxidation followed by carbonisation of oriented polyacrylonitrile filaments.

TABLE 4.12
General Types of Filamentary Carbon

Type	Method of production	Density	Tensile strength (psi × 10^5)	Young's modulus (psi × 10^6)	Fine structure	Applications
Carbon (graphite) whiskers	Special methods: e.g. formation in carbon is at high pressure	About 2·2	25–30	100–145	Graphite single crystal	No commercial applications at present
High-performance carbon fibres	Controlled pyrolysis of oriented synthetic filament precursor (normally polyacrylonitrile)	1·7–2	3–4	40–60	Fibrillar, closely determined by that of precursor filament. Fibrils consist of oriented graphite crystallites: the degree of orientation determines modulus value[48]	Reinforcement in composites, mainly thermosetting polymers (continuous carbon filament reinforcement), recently thermoplastics (mainly discontinuous fibres)
Low-performance carbon fibres	Controlled carbonisation, culminating in 'graphitisation' of a precursor fibre (normally cellulose rayon)	1–1·5	0·5–1·0	4–10[a]	Graphite layers partly oriented along fibre axis (modulus increases with degree of orientation[a]). Voids (up to 1 000 Å in size) present between the layers: these are a factor in low strength and density[40] and are reduced by hot stretching	High-temperature insulation; glands, seals and packings in chemical plant; special heating elements (graphite tape, braid and cloth).[40] Hot-stretched fibre used as reinforcement in composites

[a] Can be increased up to about ×7 by stretching at the graphitisation stage. Strength and density are also increased.

4.5.2.2 Carbon Fibre Reinforcement in Thermoplastics

The high-performance carbon fibres of interest as reinforcement in thermoplastics are of two main types: high modulus fibre ('type 1') and high strength fibre ('type 2'). Their strength and stiffness properties are shown in Table 4.1. Two additional grades, intermediate in properties between types 1 and 2, have also been produced, with the following typical mean strength and stiffness values:

Fibre type	Tensile strength (psi × 10^5)	Young's modulus (psi × 10^6)
3 (originally RAE Farnborough)	2·9	29
4 (Morganite Modmor Ltd.)	2·6–3·3	38–49

Improvements in production processes continue to be made. In particular cheaper production from ordinary, textile grade of polyacrylonitrile filaments has been disclosed, including in-line application of coupling treatments to improve interfacial adhesion to polymer matrices.[47] These features are embodied in the process developed by the Great Lakes Research Corporation (Elizabethton, Tennessee, USA), used to produce the 'Fortafil' carbon fibres.

Some suppliers of carbon fibres in the UK and USA are listed below:

UK
Morganite Modmor Ltd., London.
Courtaulds Ltd., Coventry.

USA
Union Carbide Corp., Carbon Products Div., New York, N.Y. ('Thornel')
Hercules Inc., Wilmington, Delaware.
U.S. Polymeric, Santa Ana, California.

Thermoplastic moulding compounds reinforced with carbon fibre were first introduced in 1973 in the USA by the Liquid Nitrogen Processing Corp. and Stackpole Fibres Company Inc.

4.5.2.3 Effects in Thermoplastic Polymers

An extensive recent study by Theberge *et al.*[41] of the effects of carbon fibre reinforcement in several thermoplastic polymers has shown that, in comparison with the results of reinforcement with glass fibre at corresponding loading levels, the strength was considerably improved, stiffness increased by a factor of two, and resistance to long-term flexural creep was almost doubled. Special properties conferred by the presence of the carbon fibres included electrical conductivity (at higher loading levels), thermal conductivity increased by factors of 2–3 over glass-filled analogues and by factors of 3–5 over the base polymer, reduced coefficient of friction and reduced wear (the latter especially in nylon 6.6).

4.6 PARTICULATE FILLERS

4.6.1 General Considerations

Only those particulate fillers which produce some true reinforcement effect when incorporated in a thermoplastic polymer are discussed in this section. For the purpose of the discussion they are defined as the fillers whose particles have low aspect ratios (if needle- or rod-shaped) and which may be roughly spherical (with varying degrees of irregularity) or plate-like. Strictly speaking glass spheres are a particulate filler within this definition, but because of their chemical relationship to glass fibres it was convenient to discuss them in Section 4.3.

The reinforcing effects of those particulate fillers which produce them in thermoplastics are lower both in number and in magnitude than those obtained with true fibrous reinforcing fillers, e.g. glass fibres. Thus none of the particulate fillers improves the tensile strength of thermoplastics: in fact the strength is usually lowered, especially at higher filler contents. Impact strength may also be lowered.

The property mainly improved by particulate reinforcing fillers is stiffness. Hardness can also be improved, as can the deflection temperature under load. Thermal expansion, mould shrinkage, extensibility, and creep are reduced. Surface finish may be considerably affected, and is normally better than in fibre-filled thermoplastics.

Melt properties may also be improved; e.g. talc in PVC calendering improves the hot strength of the melt, although melt viscosity is generally increased.

Two general considerations, not directly associated with mechanical reinforcement effects, are important in the use of particulate fillers in thermoplastics. These are the cost and colour of the filled compound.

Experience of the sales patterns of a variety of fillers for thermoplastics shows that—aside from some special cases where aesthetic considerations are unimportant, or where a surface coating is used on mouldings—fillers of good white colour, which make it possible to produce 'clean' coloured mouldings, are generally preferred to better reinforcing and cheaper fillers whose colour is poor.[49]

Filler cost and the effect on the cost of the compound is the second important general consideration, of special significance with particulate reinforcing fillers. Because such fillers are often used in fairly high loadings to obtain the maximum stiffness increase they are capable of giving in polyolefins, the filler cost in relation to the polymer it replaces in a given volume of compound (*cf.* numerical example in Section 4.3.3) is particularly important in deciding whether a saving can be effected (extender effect), or whether any additional cost is justified by the property improvements. Until the oil crisis the price of polyolefins (in which particulate reinforcing fillers are mainly used) was so low that incorporation of a particulate filler did not necessarily make the compound cheaper. Currently the rise in prices of polymers has brought the extender effect into its own.

4.6.2 Reinforcing Particulate Fillers in Commercial Use

4.6.2.1 TALC

Talc is a naturally occurring crystalline material, composed of hydrated magnesium and silicon oxides. The ideal composition approximates to the formula $3MgO.4SiO_2.H_2O$ but talc as mined usually contains admixtures of other minerals and impurities, not normally removed in processing, varying according to the place of origin. The physical form also varies according to the source: the particles may be needle-shaped (acicular) bordering on micro-fibrous, or plate-like (micaceous, 'platy' or scaly) or even granular. The particle size may vary, roughly between 1 and 50 μm. It is determined by the type of talc and the grinding process used in its production.

The origin, form and approximate composition of some talcs are shown in Table 4.13.

TABLE 4.13
Talcs from Various Sources

Source	Particle type	Main chemical constituents						Ignition loss (wt %)
		SiO_2	MgO	CaO	Al_2O_3	FeO + Fe_2O_3	% Purity	
USA	Platy (Western talc)	62·7	25·5	2·7	4·7	0·1	92·4	4·7
USA	Acicular (New York talc)	59·0	28·5	6·5	0·7	0·3	91·8	5·3
USA	Acicular (New York talc)	58·6	28·9	5·0	0·9	0·3	91·7	4·4
USSR	Granular/blocky	59·8	32·4	0·9	0·7	0·3	96·7	4·2
USSR	Granular/blocky	61·8	32·4	0·6	0·3	0·3	98·8	5·2
Japan	Granular/blocky	53·6	29·7	2·1	4·7	4·8	87·4	6·8
Japan	Granular/blocky	55·2	29·9	1·2	4·1	4·1	89·3	6·3

Talc is used mainly in polyolefins, especially polypropylene, and in PVC (calendered floor tile compounds). In the latter it provides some reinforcement to the hot melt, and also a stiffening effect in the finished product.[50] The main effects in polypropylene are illustrated in Fig. 4.15. Treatments (coating) of the particle surface are thought to have some effect on the properties of the compound.[51]

Talc has much less adverse effect on the thermal stability of polypropylene than asbestos[52] and the colour of the compound is better. It is a smooth, non-abrasive filler.

4.6.2.2 Wollastonite

Wollastonite is another silicaceous, mineral filler, a natural calcium metasilicate of about 97% purity after processing.[52] The particles are needle-shaped, up to about 50 μm in size. It was first used for its thixotropic effect in surface coatings and thermosetting polymer compositions.[50]

Its applications in thermoplastics are generally similar to those of talc, as are its effects (*cf.* Fig. 4.15).

It has been claimed that surface treatment of the particles with some silicone coupling agents can improve somewhat the strength (flexural and tensile) and the stiffness of polypropylene/Wollastonite compounds.[10]

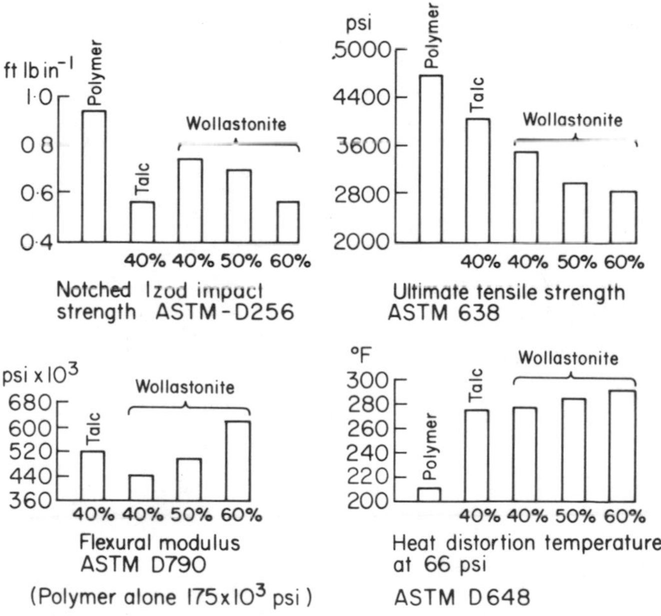

FIG. 4.15 Some effects of talc and Wollastonite in polypropylene. (Data of Fletcher and Tully: reprinted, with permission, from the Proceedings of the 23rd ANTEC, SPE, 1967.)

TABLE 4.14

Reinforcing Fillers in some Thermoplastic Polymers: Comparison of Effects on Mechanical Properties

Polymer	Filler		Properties						
	Nature	wt %	Tensile strength (psi) (ASTM D-638)	Tensile elongation (%) (ASTM D-638)	Flexural modulus (psi) (ASTM D-790)	Izod impact strength (ft lb in^{-1}) (ASTM D-256)	Heat deflection temp. at 264 psi (°C) (ASTM D-648)	Coeff. of linear thermal expansion (in in^{-1} $°F^{-1}$) (ASTM D-696)	Creep and fatigue endurance in comparison with base polymer
Nylon 6.6	None	0	11 800	60	$4·1 \times 10^5$	0·9	70	$4·5 \times 10^{-5}$	Generally improved by reinforcing fillers in some cases by factors of 2 or more (e.g. by glass and carbon fibre reinforcement).
	Glass fibre[a]	40	31 000	2–3	$16·0 \times 10^5$	2·6	260	$1·4 \times 10^{-5}$	
	Glass beads	40	14 200	2·5	$7·3 \times 10^5$	0·6	74	$2·9 \times 10^{-5}$	
	Asbestos fibre[j]	40	18 500	—	$16·0 \times 10^5$	—	—	—	
	Carbon[b] fibre	40	40 000	3–4	$34·0 \times 10^5$	1·6	260	$0·8 \times 10^{-5}$	
	Fybex[c]	40	26 000	—	$19·0 \times 10^5$	1·8	249	$1·0 \times 10^{-5}$	
Polypropylene	None	0	5 000	200–700	$2·0 \times 10^5$	0·5	60	$4·8 \times 10^{-5}$	
	Glass fibre[d]	30	9 800	2–3	$8·0 \times 10^5$	1·6	146	$1·1 \times 10^{-5}$	
	Asbestos fibre[e]	40	5 600	—	$7·3 \times 10^5$	1·3	106	$1·4 \times 10^{-5}$	
	Fybex[f]	30	7 900	—	$11·6 \times 10^5$	0·38	108	About 2×10^{-5}	
	Talc[g]	40	4 000	15	$5·0 \times 10^5$	0·5	115 at 66 psi	$2·2 \times 10^{-5}$	

ABS	None[h]	0	6 000	—	3·3 × 10^5	7·2	74	6·0 × 10^−5	As above
	Glass fibre[i]	30	14 500	3–4	11·0 × 10^5	1·4	105	1·6 × 10^−5	
	Glass beads	30	4 620	4	5·6 × 10^5	0·57	92	—	
	Asbestos fibre[j]	30	13 900	—	11·3 × 10^5	—	—	—	
	Carbon fibre[k]	16·6	6 100	—	8·4 × 10^5	—	—	—	
	Fybex[l]	20	10 200	—	8·0 × 10^5	0·50	95	2·3 × 10^−5	
Poly-sulphone	None	—	10 200	50–100	3·9 × 10^5	1·3	174	3·1 × 10^−5	
	Glass fibre	30	18 000	3	12·0 × 10^5	1·8	185	1·4 × 10^−5	
	Carbon fibre[m]	30	19 000	2–3	20·5 × 10^5	1·1	185	0·7 × 10^−5	
	Fybex[n]	25	13 500	2–3	9·0 × 10^5	0·6–1·0	180	1·5 × 10^−5	

[a] Thermocomp RF 1008 (LNP).
[b] Thermocomp RC 1006 (LNP).
[c] Linsen and Regester.[35]
[d] Thermocomp MF 1006 (LNP).
[e] Arpylene APPN 2240 (Turner Brothers Asbestos Co.).
[f] Surface-coupled.
[g] Arpylene TPPN 2040 (Turner Brothers Asbestos Co.).
[h] Medium-impact grade.
[i] Thermocomp AF 1006 (LNP).
[j] Hollingsworth.[57]
[k] Hollingsworth and Sims.[1]
[l] Medium-impact grade composite.
[m] Thermocomp GC 1006 (LNP).
[n] Thermocomp GW 1005 (LNP).

4.6.2.3 Other Particulate Fillers

Calcium carbonate (chalk, whiting) and clay, which are fairly common fillers in thermoplastics (mainly polyolefins and PVC), have some stiffening effect and reduce shrinkage and creep, but are often incorporated mainly as extenders to reduce cost.

Chemically, whiting fillers can be very pure calcium carbonate (over 99%), with crystalline particles, irregular in shape, but with a considerable degree of sphericity. The grades most suitable for use in thermoplastics have particle sizes between about 0·5 and 8 μm. Some grades are surface-treated (normally with a stearate) to aid dispersion in polymers. It is a useful rough general rule that coarse particle fillers (above 20–30 μm) can affect surface finish and reduce some physical properties, whilst very fine particles (below about 0·1 μm) affect the melt viscosity strongly.

The effects of calcium carbonate (Calibrite, Hydrocarb, Millicarb, Omya)* in polyethylene, including their use as part-replacement for TiO_2 pigments, were studied in considerable detail by Adamajtis[54] and Jangeerkahn.[55] It has been claimed that fine whiting fillers can improve not only the stiffness but also the tensile strength in some thermoplastic polymers.[56]

Clays are hydrated alumino-silicate minerals. The particles of clay fillers used in thermoplastics (which may be calcined or uncalcined ground clay) are platelets. The particle size is comparable to that of whiting (0·2–8 μm). The effect on mechanical properties is also generally similar, but the colour is comparatively poor and the particles are more abrasive.

Recently clay fillers have been incorporated in nylon, mainly as extenders in view of current polymer shortage, although some improvements in stiffness are being claimed (see also Chapter 8, Section 8.2).

4.7 COMPARISON OF THE REINFORCING EFFECTS OF SOME FILLERS IN FOUR THERMOPLASTICS

Table 4.14 shows a comparison of some of the effects of fillers on properties of two 'engineering' thermoplastics (nylon and polysulphone) and two widely used commodity plastics (polypropylene and ABS) which can be upgraded to the engineering class by suitable reinforcement.

REFERENCES

1. Hollingsworth, B. L. and Sims, D. (1969). *Composites,* **1,** 2, 80.
2. Hartley, P. (1971). *Design and Components in Engineering,* January 20th, p. 70.

* Plastichem Ltd., Esher, England.

3. Hodgson, A. A. (1965). Fibrous Silicates, Lecture Series No. 4, Royal Institute of Chemistry, London.
4. Mark, H. F. (Ed.) (1967). *Encyclopedia of Polymer Science and Technology,* Vol. **6**, Interscience Publishers, New York.
5. Ainsworth, L. (1971). *Composites,* **2,** 1, 14.
6. Cook, J. G. (1964). *Handbook of Textile Fibres,* 3rd edn, Merrow Publishers Co., Watford, England, p. 388.
7. Maaghul, J. (1971). 26th Conference, RP/C Division, SPI, Paper 8-A.
8. Richards, R. W. and Sims, D. (1971). *Composites,* **2,** 4, 214.
9. Crabtree, J. D. and Pickthall, D., Paper presented at the IRI Conference on 'Polymer Blends and Reinforcement', September, 1969, Loughborough, England.
10. Plueddemann, E. P. (1972). 27th Conference, RP/C Institute, SPI, Paper 11-B.
11. Cameron, G. M. and Marsden, J. G. (1972). *Chemistry in Britain,* **8,** 9, 381.
12. Plueddemann, E. P. (1965). 20th Conference, SPI, Paper 19-A.
13. Plueddemann, E. P. (1967). US Patent 3 306 800.
14. Zisman, W. A. (1963). *Ind. Eng. Chem.,* **55,** 19.
15. Hartlein, R. C. (1971). *Ind. Eng. Chem. Prod. Res. Develop.,* **10,** 1, 92.
16. Anon. (1971). *Composites,* **2,** 1, 7.
17. (1972). *Ibid.,* **3,** 3, 100.
18. Raask, E. (1968). *J. Inst. Fuel,* **15,** 339.
19. Strauch, O. R. (1969). *SPE Journal,* **25,** 9, 38.
20. Hopkins, R. C. (1972). *P.R.T. Polymer Age,* **3,** 9, 344; Anon. (1969). *Plastics Design and Processing,* September, p. 30.
21. Ritter, J. (1970). 25th Conference, RP/C Division, SPI, Paper 8-A.
22. Anon. (1968). 'Successful "Zero Reject" Program', *Plastics World,* December; Wotitzky, H. J. (1969). SPE RETEC, *Reinforced Thermoplastics, Properties, Processes and Uses,* Hartford, Conn., October 15th.
23. Anon. (1973). *Europlastics,* **11,** 46, 60.
24. 'Amphibole Asbestos', Technical publication of Cape Asbestos Fibres Ltd., 1973.
25. Wicker, G. L. (1971). *Composites,* **2,** 4, 221.
26. British Patent No. 945 202 (Avisun Corp.), published December 23, 1963.
27. British Patent No, 1 009 783 (Haveg Industries Inc.), published November 10, 1965.
28. Parks, L. F. Cape Asbestos Fibres Ltd., Research Centre, private communication.
29. Harrison, P. and Sheppard, R. F. (1970). 'Reinforced Polypropylene—Properties and Prospects', Paper presented at the PI Conference on Reinforced Thermoplastics, October, Solihull, England.
30. Gilson, J. C. (1972). *Composites,* **3,** 2, 57.
31. Holmes, S. (1972). *Ibid.,* 60.
32. Hollingsworth, B. L., Sims, D., Ledbury, K. J. and Brokenbrow, B. E. (1969). 24th Conference, RP/C Division, SPI, paper 1-A.
33. Richards, R. W. and Sims, D. (1970). 'Reinforcement of Thermoplastics with High-modulus Fibres', Paper presented at the PI Conference on Reinforced Thermoplastics, October, Solihull, England.
34. Mohr, J. G. (Ed.) (1973). *SPI Handbook of Technology and Engineering of RP/C,* 2nd edn, Van Nostrand Reinhold Co.

35. Linsen, P. G. and Regester, R. F. (1972). 27th Conference, RP/C Institute, SPI, Paper 11-D.
36. 'Reinforcement of Thermoplastics with Fybex Inorganic Titanate Fibres', Bulletin No. 1, DuPont Inorganic Fibres Division, November, 1971.
37. 'Fabrication and Electroplating of Fybex Fibre Reinforced Thermoplastics', Bulletin No. 2, DuPont Inorganic Fibers Division, November, 1971.
38. Weston, N. E. (1972). *SPE Journal*, **28,** 12, 37.
39. Phillips, L. N. (1967). *Trans. J. Plast. Inst.*, August, 589.
40. Gill, R. M. (1972). *Carbon Fibres in Composite Materials*, The Plastics Institute and Iliffe Books, London.
41. Theberge, J., Arkles, B. and Robinson, R. (1974). 29th Conference, RP/C Institute SPI.
42 Tang, M. M. and Bacon, R. (1964). *Carbon*, **2,** 3, 211.
43. Bacon, R. and Shalamon, W. A. (1967). 8th Conference on Carbon, State University of New York, Buffalo, USA, June, Paper M1-58.
44. Shindo, A. (1961). Report No. 317, Govt. Ind. Res. Institute, Osaka.
45. Standage, A. E. and Prescott, R. (1966). *Nature*, **211,** 5045, 169.
46. Hawthorne, H. M. (1971). International Conference on Carbon Fibres, The Plastics Institute, London, February, Paper 13.
47. Prescott, R., Goan, J. C., Hill, J. E., Joo, L. A. and Martin, T. W., (1972). 27th Conference, RP/C Institute SPI, Paper 13-C
48. Blakelock, H. D. and Lovell, D. R. (1969). 24th Conference, RP/C Division SPI, Paper 6-B.
49. Private communication from Mr. R. C. Hopkins, Marketing Manager, Plastichem Ltd., England.
50. Hunt, R. E. (1969). *Plastics Technology*, **15,** 11, 38.
51. British Patent 1 073 804 (Sumimoto Chemical Co.), published June 28, 1967.
52. Fletcher, W. J. and Tully P. R. (1967). Proceedings 23rd ANTEC SPE, 537.
53. Jones, R. F. *ibid.*, p. 533.
54. Adamajtis, S. (1974). The effect of calciferous fillers on the properties of low density polyethylene, Project Report, Polytechnic of the South Bank, London.
55. Jangeerkahn, A. (1974). Partial replacement of titanium dioxide in low density polyethylene by various forms of calcium carbonate, Project Report, Polytechnic of the South Bank, London.
56. Melbourn Chemicals Ltd. Technical Bulletin 3S3/70, May 1974.
57. Hollingsworth, B. L. (1969). *Composites*, **1,** 1, 28.

CHAPTER 5

THE MANUFACTURE OF REINFORCED THERMOPLASTIC MATERIALS

5.1 THE BASIS AND DEVELOPMENT OF METHODS IN CURRENT PRACTICE

Some indication has already been given in the Introduction, Section 1.1, of the origin and development of the main methods of production of reinforced thermoplastics.

Whilst the full details of the processes used to produce the commercially available reinforced thermoplastics are not revealed by manufacturers, the principles of the techniques employed are well known. The long established methods of the plastics industry for the incorporation of fillers and additives into polymers—largely based on melt compounding—were applied to incorporating glass fibres in polyamides as early as 1946 (British Patent 618,094). However, for a fairly long period thereafter the compounding of glass fibres into thermoplastics was not regarded as a useful practical means of producing reinforced moulding materials. The Fiberfil/Bradt coaxial extrusion method initially held sway in the United States and, with patent protection afforded by US Patent 2,877,501, other manufacturers were not active. The early 1960s saw a resurgence and spread of the use of melt compounding procedures for the incorporation of glass fibres into thermoplastics, practised particularly by LNP in the USA, ICI in the United Kingdom, and Bayer in Germany.

The two alternative general principles—coaxial extrusion and melt compounding—still underlie virtually all present-day manufacturing practice. In the following sections their more important practical embodiments are outlined, including the already-mentioned Fiberfil/Bradt process (which represents the sole commercially significant application of coaxial extrusion) and the main variants of the melt compounding route, i.e. melt extrusion and internal mixing, the latter being of some importance in the incorporation of asbestos fibre into thermoplastics. Some methods based on other principles, less important commercially, are also described.

5.2 COMMERCIAL PROCESSES

5.2.1 Coaxial Extrusion

In this method continuous glass rovings (commonly 6–8) are passed through the cross-head die of an extruder where the polymer to be reinforced is extruded around each roving; the coated rovings emerge from the die consolidated into a single strand. This is cooled and then cut to form granules.

The extruder design is of secondary consequence in this technique as clearly it is not required to process a glass fibre/polymer mixture. The screw geometry may be such as to be universally applicable to a wide range of polymers or may be optimised for specific polymers or polymer. Several types of cross-head die have been described, mainly in Patent Specifications.

Tensioning devices are required to ensure that the roving runs smoothly through the die into the pelletiser.

The pellets produced by coaxial extrusion have the structure shown schematically in Fig. 5.1. It will be noted that the fibres extend the full length of the granules and, like the granule, are approximately $\frac{3}{8}$ in in length. However, they are not uniformly and individually coated with the polymer.

Figure 5.2 shows electron scan micrographs of part of a section through a coaxial extruded pellet (glass-fibre-reinforced nylon 6.6).

Moulding compounds with the coaxial pellet configuration are the main type of reinforced thermoplastic material supplied by the Fiberfil Division of Dart Industries Inc. in the United States.

5.2.2 Melt Compounding

The practical embodiments of this principle are far more numerous than those of coaxial extrusion and the number of companies using this

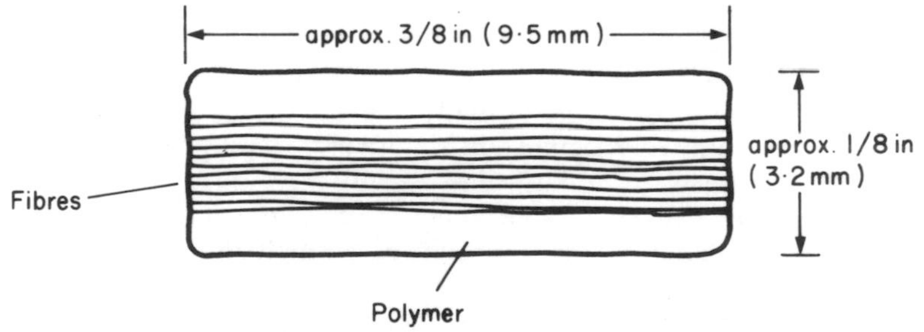

FIG. 5.1 Coaxial extruded pellet, schematic representation.

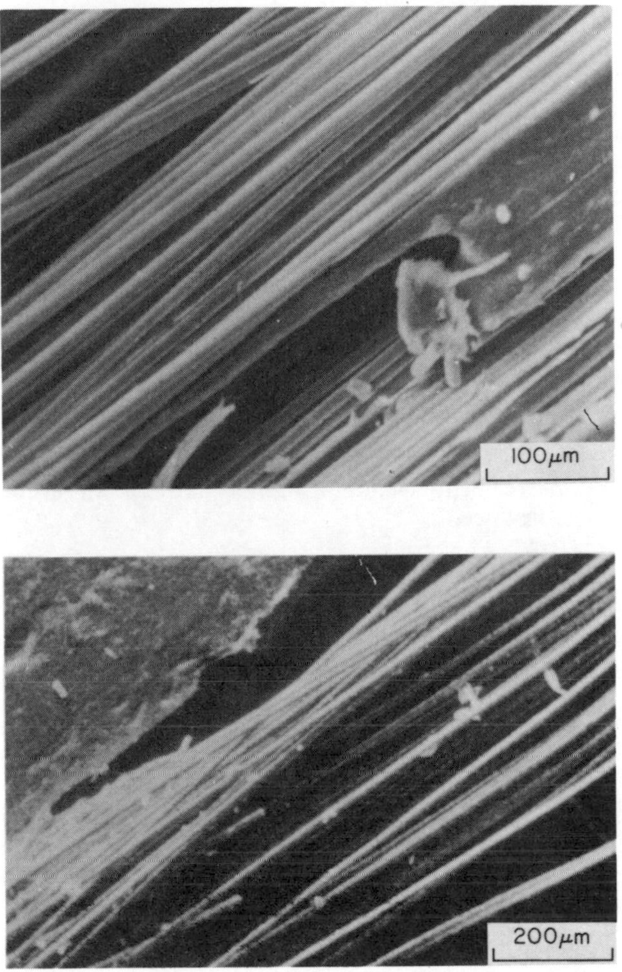

FIG. 5.2 Section through coaxially extruded pellet: electron scan.

process world-wide is also much greater. The commercial processes may conveniently be considered under the two headings of melt extrusion and internal mixing. The former group is by far the more important and more widely practised commercially. Internal mixing is normally resorted to where bulky fillers and reinforcements, e.g. asbestos, are to be incorporated.

5.2.2.1 Extrusion Compounding

The cardinal features of this type of process are that the equipment used is an extruder in which, at some stage, the reinforcement fibres are dispersed in essentially random fashion in the molten polymer and remain in such substantially random dispersion in the compound produced. There is invariably some breakdown of fibres in processing and the ultimate fibre length in the compound depends on the conditions (governed by type of equipment used and the nature of the polymer being compounded). Some effects of differences in processing conditions are illustrated in Fig. 4.2 (Chapter 4, Section 4.2.4) by glass fibres recovered from extrusion-compounded nylon and polypropylene compositions. The fibre distribution in a glass-reinforced nylon compound produced by extrusion is shown in Fig. 5.3. The fibre loading is comparable with that in the parallel-fibre pellet of Fig. 5.2.

The main versions of the extrusion compounding process differ in the type of extruder used: there are also differences in the form in which the fibrous reinforcement is fed in, as well as in the point on the extruder at which it is introduced.

5.2.2.1.1 single-screw extrusion compounding

The use of a single-screw extruder for the incorporation of glass fibres into a polymer entails modification of conventional machines to minimise the adverse effects of the strong, hard, abrasive material (the glass) which is being introduced. Glass is the reinforcement normally processed, as single-screw machines are seldom used for reinforcing polymers with other fibres.

General description

Machines currently used by the major manufacturers of reinforced thermoplastics materials are two-stage vented extruders incorporating very powerful drives and fitted with specially modified die heads. The heart of the extruder, the screw, can, as mentioned above, be optimised for specific polymers or may be a general-purpose one suitable for a number of polymers. The details of the design of special screws are often jealously guarded commercial secrets. However, they all must have deep feed flight sections and comparatively high compression ratios to allow for the conversion of the bulk feed to a homogeneous melt. Venting is a standard feature of the equipment, to remove volatiles arising in the polymer melt (e.g. traces of moisture or monomers) or from size on the fibres (e.g. plasticiser vapours or products of thermal decomposition).

The extrusion of a mixture of glass and plastics melt requires considerably more power per pound of output than is called for with unfilled or lightly filled materials. Consequently very high horse-power drives are used on machines of comparatively small screw diameter. Some

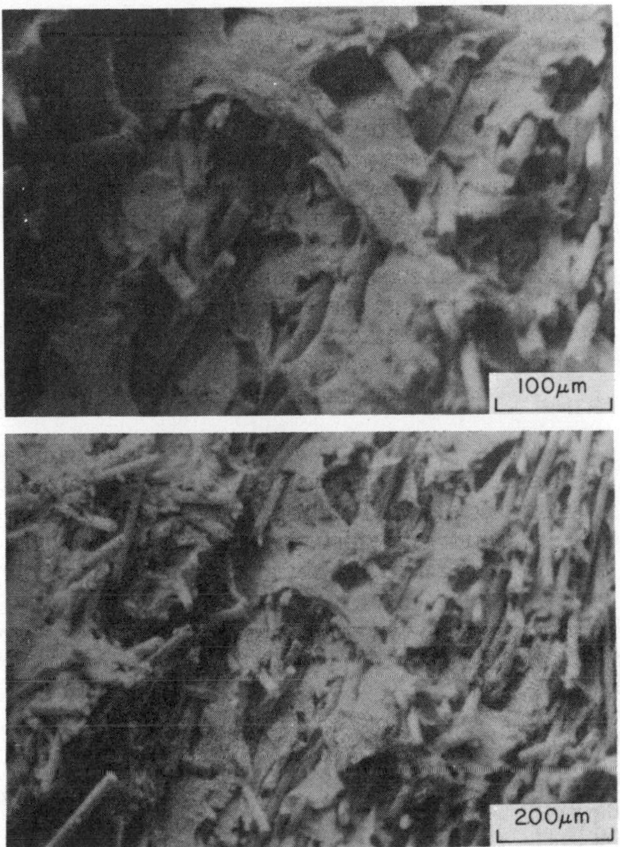

FIG. 5.3 Electron scan micrographs of a section through a pellet of nylon/glass-fibre composition, extrusion-compounded.

manufacturers aim at a material in which the reinforcing glass fibres are comparatively short, giving good dispersion and a good surface finish in the ultimate mouldings but less-than-optimum physical properties. To produce such compounds the amount of shear imparted in the extruder is high, usually achieved by high screw speeds which also result in high outputs. Other producers elect to make compounds in which the fibres are not extensively broken down in the course of mixing and are consequently comparatively long. Mouldings produced from such compounds usually have better mechanical properties (*cf.* Chapters 1 and 6) but the surface finish may be somewhat inferior. The degree of breakdown of the fibrous filler (the final fibre length distribution and average fibre length) is strongly influenced by the design of the screw. For a given screw geometry it is generally dependent on the amount of shear experienced by the melt and the residence time, and is normally a function of the screw speed and the head pressure. The method and point of introduction of the feed can also play a part.

Resistance to wear through abrasion by glass fibres is an important factor in the selection of construction material for the screw and barrel. However, each manufacturer chooses between low cost frequent replacement and high cost with a lower replacement rate. In the case of screws the choice is usually between a nitrided steel and a special hard alloy coating covering not only the flights but the entire root of the screw. The most usual candidate for barrel construction is the bi-metallic type exemplified by Xaloy (United States and Belgium), Brux (UK) and Bernex (Switzerland).

The die-head must be designed or modified to allow for the extrusion of the reinforced material without blockage by matting of the reinforcing fibres. Wear in this section of the machine is not so serious as the material is fully plasticised at this point.

Feed arrangements

The normal point of introduction of the polymer and fibres is conventionally located, i.e. a feed throat in the usual position at the back of the barrel. Down-stream feeding, with the fibres introduced at a point part way down the barrel, i.e. into pre-fluxed polymer, is not usual. Fibres introduced at that point would not be readily taken up and dispersed in the melt by the action of a single screw.

Three possible variants of feed procedure are:

(1) feeding a pre-mix of polymer granules and chopped fibre;
(2) separate introduction into the feed throat of the polymer and the chopped fibres;
(3) normal polymer feed with the fibres introduced into the feed throat as a continuous strand which is torn up as it is taken up by the screw.

All three variants are represented in patent specifications: in practice the first is the most popular and the third is not normally employed.

Take off

The conventional methods of taking off and granulating the extrudate can be employed with reinforced thermoplastics materials. They are:

(1) *Strand extrusion.* In this the strands, typically 20–30 in number, are extruded in line, pulled away from the die-face by feed rollers and passed through a water bath or an air cooling system to the pelletiser. This system requires more floor space than the alternative method of die-face pelletising.

(2) *Die-face pelletising*. In this method the holes of the die are usually arranged in a circle and the emerging extrudate is cut by a spinning knife at the die-face itself. The pellets so formed are either drawn into some form of cyclone separator, or into a stream of cold air or water, from which they are subsequently separated by screening.

The wear on cutting knives is very much higher than that experienced with ordinary thermoplastics and careful selection of knives and steels is important.

A general point is worth making, concerning the scope of the two take-off systems in their application to reinforced thermoplastics. Strand extrusion, involving as it does the cutting of solidified, cool material, is widely applicable, largely irrespective of the nature of the polymer and reinforcement. The effectiveness, and hence the applicability, of die-face pelletising on the other hand, with its essential feature of cutting (shearing) a semi-molten filled polymer composition, does depend on the nature of the polymer and also, in certain cases, the nature and level of reinforcement because these factors influence the shear properties of the material. Experience shows that equally good results over a broad spectrum of polymer/reinforcement combinations cannot be assumed *a priori*. However, where a limited range of polymers and/or compositions is processed, the manufacturing experience will soon establish whether die-face cutting is practicable.

Application

Single-screw extruders are widely used for the manufacture of glass-reinforced nylon and are also employed by a number of companies producing a wide range of reinforced thermoplastics. Their advantages and disadvantages in comparison with twin-screw extruders are shown in Table 5.3 below.

5.2.2.1.2 TWIN-SCREW EXTRUSION COMPOUNDING

The commercial use of twin-screw compounding for reinforced thermoplastics materials in the main post-dates that of the single-screw method. However, because of the advantages of the process and the highly effective modern equipment available, much of the recent commercial activity has centred on the use of twin-screw machines.

The growth in popularity of the twin-screw extruder in the plastics industry has been largely due to its ability to process PVC, and in particular rigid PVC, from powder blends. A machine in which two meshing screws can effect positive pump action is also capable of more efficiently compounding and extruding bulky feed stock than a single-screw extruder. Moreover, single-screw extruders rely to a large extent upon the development of high head pressures to produce a homogeneous melt whereas the twin-screw extruder can render polymers fully plasticised almost entirely by the action of the screws. The dispersing action when compounding-in a second phase (the reinforcement, usually fibrous) is also better in the twin-screw machines.

As a result of its increasing popularity, and in particular its growing field of application in PVC compounding and extrusion, several makes of the twin-screw extruder have appeared on the market. Most of these, whilst they have good compounding action, are also conceived and employed as extruders for the production of semi-finished goods, i.e. pipe and profiles. Most would probably be suitable, possibly with modifications, for the production of reinforced thermoplastics materials by melt compounding. It is therefore not necessary to consider in detail all the available types. Instead, a typical, good twin-screw extruder with compounding action, which is used *inter alia* for the commercial production of reinforced thermoplastics materials, will be briefly described and compared with a specialised twin-screw machine designed primarily for compounding action.

The two types of machine are represented respectively by the Mapre* extruder and the Werner Pfleiderer compounder-extruder†.

The Mapre extruder

This is a twin-screw machine with typical intermeshing co-rotating screws (L/D ratios 12:1, 14:1 or 16:1) driven by an infinitely variable hydraulic motor. The general range includes several screw design modifications. Special designs are also available, e.g. with de-gassing and high-intensity mixing sections.

In twin-screw machines the speed of rotation is slower than with single-screw extruders used for melt-compounding reinforced thermoplastic materials: typically it is about 40 rpm. Table 5.1 gives some details of two Mapre 100 mm screw extruders. Typically these can provide output rates of about 250 kg h^{-1} in average production of reinforced thermoplastics. However some materials can be run faster than others and the specific gravity of the composite will also affect the weight output rate for a given volume. Thus, for example, the output in weight terms of a 20% glass-filled polypropylene (specific gravity about 1·04) will be considerably lower than that of a 40% glass-reinforced polyacetal (specific gravity 1·71), extruded in the same volume by the same machine. A larger model (160 mm screw) is also available.

As can be seen in Fig. 5.4, with standard Mapre screws compression is obtained by increasing the thickness of the flights towards the exit end of the barrel. The photograph also shows modification of the flights in one section. The screws may be nitrided steel, or may have the flight

* Nouvelle Mapre S.A., Luxembourg.

† Werner and Pfleiderer, Stuttgart, West Germany.

TABLE 5.1
Outline Specification of a Mapre Extruder
(Courtesy of D. Dryburgh & Co. Ltd.)

Technical data	Model E2 65 100-12D	Model E2 66 100-16D
Screw diameter (mm)	100	100
Stepless screw speed range (rpm)		
(a) Hydraulic drive	0–50	0–50
(b) PIV drive	5–30 or 10–60	5–30 or 10–60
(c) Commutator motor drive	5–30 or 10–60	5–30 or 10–60
(d) Direct current drive	5–50	5–50
Approx. output rates (kg h^{-1})		
(a) For tubes and sections	50–120	50–200
(b) For pellet production	100–300	100–360
Power of driving motor (hp)		
(a) Hydraulic drive, PIV drive	40	40
(b) Commutator motor drive, direct current drive	38	38
Heating capacity of barrel (kW)	min. 14	min. 22
Air cooling (m^3 h^{-1})	4 × 600	4 × 600
Thrust bearings for back-pressure (kg cm^{-2})	800	800
Net weight approx. (kg)	4 000	4 500

Special features. The thrust box is a new design, machined from solid steel block to eliminate possibility of movement of the bearing assembly. The load-sustaining quality is trebled in comparison with previous design.

Barrel heating is by cartridge heaters, allowing quick replacement in case of failure. Air-cooling is used in conjunction with a cowling for better coolant contact. The barrel can be moved forward and rotated through 90° for cleaning and general access (Fig. 5.5).

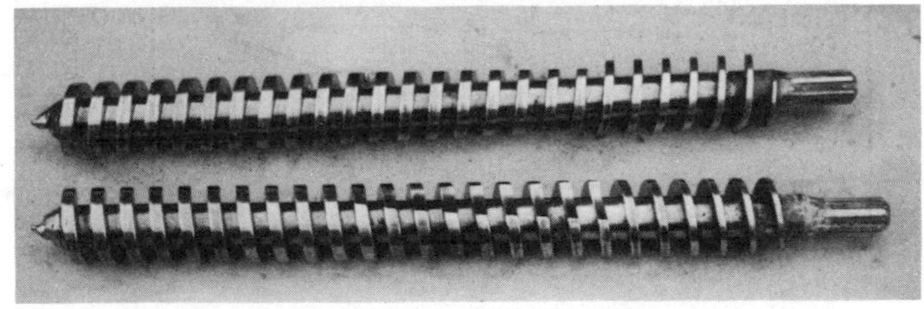

FIG. 5.4 Mapre screws. (Courtesy of D. Dryburgh and Co. Ltd.)

tips coated with a hard alloy (typically Stellite). They may be fitted with an internal heat transfer system (front-to-rear).

In common with twin-screw machines used for the extrusion of unreinforced materials, the power requirement of a compounding twin-screw extruder is lower than that of a single-screw machine (*cf.* the horse-power rating of the hydraulic drive of the Mapre model in Table 5.1).

Typically for equipment of this kind the standard Mapre machine recommended for the production of reinforced thermoplastics materials is fitted with a hardened steel barrel.

The prospective user of twin-screw extruders is currently acquiring the choice of a bi-metallic barrel, with its greater resistance to wear. Until recently bi-metallic twin-screw barrels have not been available due to technical problems in centrifugally casting the hard face-alloy. Such barrels are now appearing as a result of recent developments but they are not yet generally available. It is claimed that the need for bi-metallic coating is not as great with twin-screw machines as in the single-screw machines because of the lower screw speeds. Justification for this claim in relation to any particular model can only be obtained from practical experience of continuous production. Some Mapre extruders, with nitrided screws and barrels, used for compounding modified PPO with glass fibres, are claimed to have operated for up to two years before replacements became necessary.

FIG. 5.5 Mapre extruder with barrel swung round for cleaning. (Courtesy of D. Dryburgh and Co. Ltd.)

It may be noted in passing that Mapre twin-screw machines for processing *thermosetting* materials have included split barrels, with replaceable linings, for several years past. These features would be of interest to the compounder of glass-reinforced thermoplastics if they could be engineered into the twin-screw thermoplastics compounding extruders. The Buss Ko-Kneader which has a split barrel is a single-screw, round-barrel special compounding machine (*cf.* Section 5.2.2.1.3).

It is an important general point that operators of twin-screw extruders for compounding reinforced thermoplastics materials have the choice of introducing the normally abrasive fillers either in the conventional way at the feed pocket or down-stream, part way down the barrel. In the second arrangement the abrasive filler may not only never come in contact with the root and back portion of the screw, but will, moreover, be fed into the already fairly well plasticised polymer: this reduces the abrasive effect on those parts of the screw and barrel which experience contact with the filler. For these reasons there is less need to cover the flights and the entire length of the screws in a twin-screw machine with a hard coating.

The considerations governing the design or modification of the die-head are basically the same as those applicable in single-screw extruders for the compounding of reinforced thermoplastics materials. Similar feed systems are employed on both the general compounder/extruder machines of the Mapre type and the specialised compounding machines like the Werner Pfleiderer. The typical feed systems for twin-screw extruders are therefore discussed after the general description of the Werner Pfleiderer compounder.

The Werner Pfleiderer compounding extruder

Figure 5.6 is a general view of the ZSK 83 model (83 mm screw).

The Werner Pfleiderer ZSK extruders typify the specialised compounding twin-screw machine. They are conceived primarily as advanced compounding plant and are not normally used for the production of extruded products like profiles, pipe, etc. The design and construction of the compounding extruders are focussed on the effective inter-mixing of the polymer with the additives.

The equipment uses twin inter-meshing, co-rotating screws: the design ensures that the root of each screw is wiped by the flight tip of the adjacent one. This self-cleaning profile eliminates dead spots and normalises the residence time of the material. Along the length of the screw sections of this type of configuration may be interrupted by the incorporation of special triangular kneading blocks. As illustrated in Fig. 5.7, the screws may be built up from a number of sections to form the specific configuration required by the compounder.

Venting ports can be provided at several points along the barrel.

FIG. 5.6 Werner-Pfleiderer Compounding Extruder Model ZSK 83. (Courtesy of Werner and Pfleiderer, via Baker Perkins Ltd.)

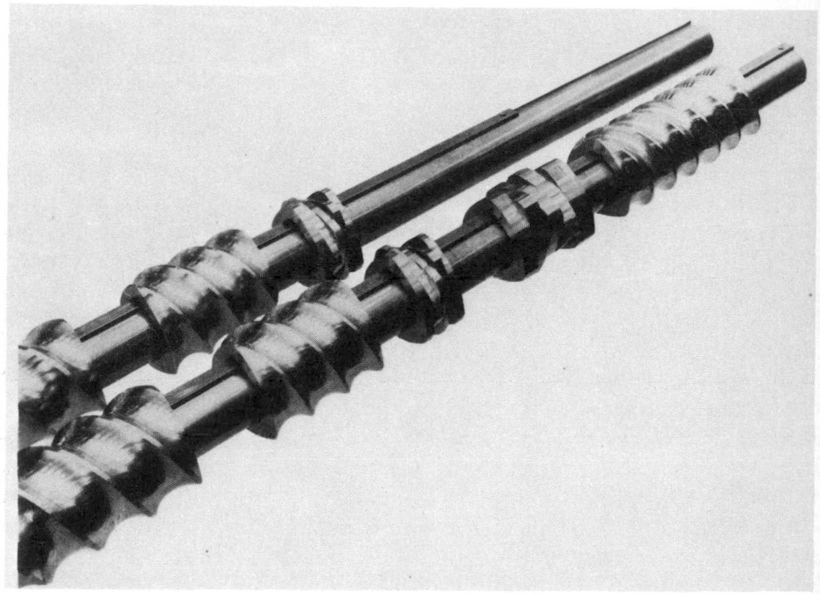

FIG. 5.7 Extruder screw shafts with some screw bushings assembled. (Courtesy of Werner and Pfleiderer, via Baker Perkins Ltd.)

The main processing features offered by the design are summarised by the makers as follows:

(1) positive conveyance of material with adjustable residence time distribution;
(2) controlled build-up of pressure at specified locations along the barrel by use of reverse-flight screw bushings, or by adjustable dynamic valves;
(3) controlled shear rate profiles by variable configuration of kneading zones, selectively combined with dynamic valves and with the temperature control of barrels and screw;
(4) heat exchange systems with accurately controlled, high intensity heating and cooling capacities;
(5) material exchange processes at relatively high velocities due to thin material layers, continuous renewal of surfaces and inversion of the material stock.

The operational advantages of this type of equipment were discussed quite early in the literature.[1]

The machine as supplied for compounding reinforced thermoplastic materials is essentially the same as that intended for non-reinforced thermoplastics but the metals selected for both the barrel and the screws are of the hard alloy type.

The ZSK machine with screw diameters of 83 mm (approximately three inches) can be expected to give an approximate output rate of 200 kg h^{-1} depending upon the polymer and reinforcement being extruded. The screw speed at this output would be between 150 and 200 rpm, and the recommended motor horse power 100. The weight output is subject to the same general consideration which applies to any other compounding extruder, *viz.* that—at least for moderate levels of reinforcement—the weight output will increase with increasing specific gravity of the polymer and the reinforcing filler.

An outline specification for a typical Werner Pfleiderer machine suitable for compounding reinforced thermoplastics is given in Table 5.2. Both larger and smaller units are also produced (screw diameters 53, 120 and 160 mm).

An important feature of the Werner Pfleiderer machines is that they are offered complete with proprietary systems for feeding either continuous or discontinuous glass fibres (roving or chopped strand) to the extruder at a location on the barrel downstream from the melting zone.

Downstream feeding is normal practice with twin-screw extruders for compounding reinforced thermoplastic materials: it is considered separately in the following section.

It is relevant to note that various aspects of the use of Werner Pfleiderer ZSK machines for the production of thermoplastics reinforced with glass fibre are covered by several patents and patent applications.

TABLE 5.2
Outline Specification for the Werner Pfleiderer Reinforced Thermoplastics Compounding Unit ZSK 83/v
(Courtesy of Werner & Pfleiderer via Baker Perkins Ltd.)

Technical data	Units	Values	Remarks
Screw diameter	mm	83	The cooling capacity is designed for each process and for the specific use of the machines, in view of the desired temperature profile along the process section, and depending on the locally available cooling facilities
Maximum screw length[a]	mm	3 150	
Depth of screw threads	mm	7·5	
Screw rotation	rpm	150–200	
Maximum torque per screw shaft	ft lb	1 700	
Main drive power	kW	75–100	
Electric heater capacity			
for $L/D = 30 = 10$ barrel sections	kW	60	
for pelletiser and screen changer	kW	20	
Throughput rate (screw speed 150 rpm)			Specific energy input (kWh kg^{-1})
Compounding polypropylene with glass fibres (approx. 30%)	kg h^{-1}	200–300	0·15–0·25
Compound nylon with glass fibres (approx. 30%)	kg h^{-1}	180–250	0·15–0·25

[a] Depending on the motor type.

The patent position should therefore be considered by the prospective user.

Feed arrangement in twin-screw compounder extruders

The feed arrangements constitute one of the important points of difference in compounding practice with single-screw and twin-screw extruders. It is a considerable advantage of the latter that they are suitable for downstream feeding in which, as has been mentioned, the reinforcement (commonly the hard, abrasive fibres) is introduced at a point part-way down the barrel, into polymer already pre-fluxed. This reduces the amount of working undergone by the fibres and the frictional heat generated, which is also less because of the lubrication provided by the polymer melt. At the same time dispersion of the fibres in the liquid melt makes for better overall efficiency of mixing. These factors cut down fibre breakdown as well as wear on the working parts (screw and barrel lining). The wear is also confined essentially to the front section of the extruder where it is again reduced by the lubricating action of the melt.

Possibly the earliest statement of the downstream feed principle, its advantages and typical general arrangement, is contained in the Specification of the Dow British Patent No. 1 087 859 (and the corresponding American Patent Application whose priority it claims).

Apart from the location of the feed point, the second main general feature of the feed is the form and order in which the polymer and fibrous reinforcement are introduced into the extruder. Reference has been made to this in connection with feed arrangements in single-screw extruders where, because the single-screw action is not as efficient as that of twin screws in disintegrating continuous rovings, feeding of fibres in that form is not widely practised.

In twin-screw extruders with their much more positive and forceful shearing action, feeding of continuous glass-fibre rovings in conjunction with accurately metered polymer is fairly normal practice. Figure 5.8

FIG. 5.8 Continuous glass-fibre strand entering the feed port of a twin-screw compounder extruder. (Experimental set-up: Courtesy of Werner and Pfleiderer, via Baker Perkins Ltd.)

shows glass rovings being taken up by the screws in the feed port of a Werner Pfleiderer machine. It should be noted that the set-up shown is experimental and the feed port is not of the purpose-designed type, standard for production machines.

The two other possible ways of feeding the components, i.e. as a pre-mix of polymer granules and chopped fibre, or polymer granules and chopped fibres separately, are also acceptable. Thus in this respect the twin-screw machines are more versatile and effective than single-screw extruders.

The feed variant involving downstream feeding of continuous rovings is more suitable for long production runs of a single compound: the losses that can be incurred on short runs in achieving steady running conditions with such an arrangement may be unacceptable to a commercial operator.

Take-off

Of the two common methods of take-off and granulation, i.e. strand-extrusion and die-face pelletising (*cf.* Section 5.2.2.1.1 and Fig. 5.9) the former is generally preferable for use with both the extruders discussed above as examples of good twin-screw machines suitable for the production of reinforced thermoplastic materials. Neither of their manufacturers currently recommends the use of die-face pelletising as being universally applicable to a wide range of reinforced thermoplastics. This is because of the potential limitations which have been mentioned in connection with die-face pelletising in single-screw extrusion (*cf.* Section 5.2.2.1.1).

The feed and take-off arrangements discussed in the last two subsections are schematically illustrated in Figs. 5.9, 5.10 and 5.11.

5.2.2.1.3 SPECIAL SCREW-TYPE MACHINES

Under this heading may be grouped those screw-type compounding machines which are neither single-screw nor twin-screw extruders but which are or may be used for melt-compounding base polymers with reinforcing fillers to produce reinforced thermoplastics. Several such machines are available, for example the Buss Ko-Kneader* (single-screw in an internally toothed barrel), the Schalker† (planetary multi-roller) compounder, or the Welex‡ compounder (two non-meshing screws). Apart from the Buss Ko-Kneader, machines in this category are not widely used for the production of reinforced thermoplastic materials.

In the Buss Ko-Kneader the compounding action is provided by a

* Buss Ltd., Basle, Switzerland.

† Gerwerkschaft Schalker Eisenhutte, Gelsenkirchen, W. Germany.

‡ Welex Inc., King of Prussia, Pa., USA.

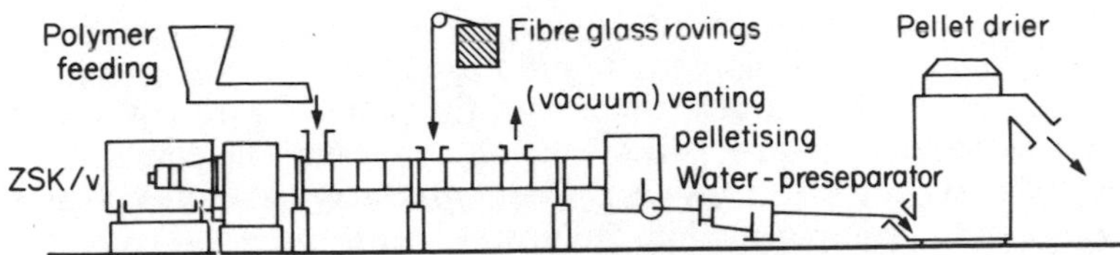

FIG. 5.9 Continuous glass-fibre roving, fed (downstream) in conjunction with metered polymer. Pelletising by hot die-face cutting, schematic representation. (Courtesy of Werner and Pfleiderer, via Baker Perkins Ltd.)

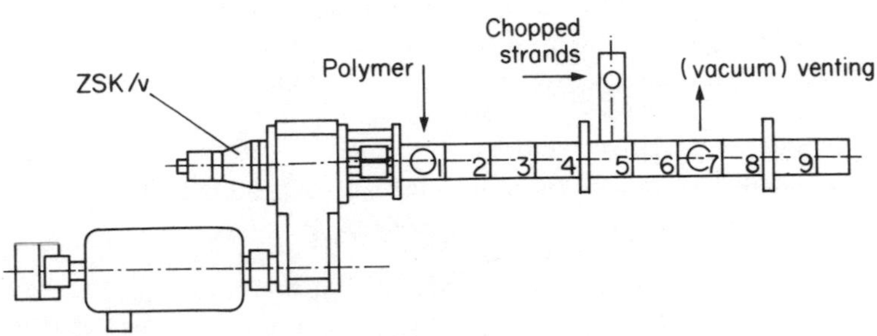

FIG. 5.10 Separate feed of polymer (feed throat) and chopped glass strands (downstream, via side-mounted single-screw extruder), schematic representation (plan view). (Courtesy of Werner and Pfleiderer, via Baker Perkins Ltd.)

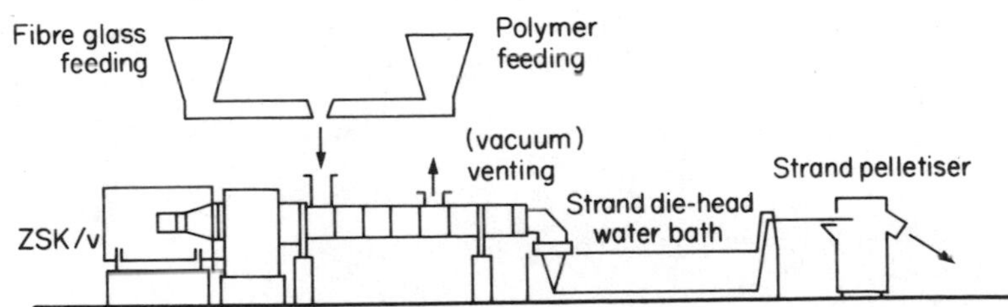

FIG. 5.11 Simultaneous, separate, metered feeding of polymer and chopped glass strands. Strand take-off with remote granulator, schematic representation. (Courtesy of Werner and Pfleiderer, via Baker Perkins Ltd.)

screw with partly discontinuous flights, rotating and reciprocating in a toothed barrel in which the teeth interact with the screw flights. In the set-up suitable for compounding reinforced thermoplastics, the pulsating extrudate from the reciprocating screw is fed to a short, conventional, single-screw extruder positioned at right angles to the main barrel. This extruder irons out the pulsations and feeds the melt continuously to the die-head. Like the compounding extruders described previously, the Buss

Ko-Kneader can be fed by the conventional hopper feed pocket with both reinforcement and polymer in pre-mixed form, or fibrous reinforcement can be metered in downstream of the point of melt formation in the barrel. Metering hoppers, dies, take-off and pelletising equipment can be supplied as part of the system. The barrel may be vented. It is also a feature of the equipment that the barrel is constructed in two halves, which may be opened out along the whole length for access to the interior and the screw.

5.2.2.2 Comparison of Extrusion Equipment and Processes

Some important general advantages and limitations of the equipment and processes in single-screw and twin-screw extrusion compounding of reinforced thermoplastics are compared in Table 5.3. Special screw-type compounding machines (*cf.* Section 5.2.2.1.3) are similar to twin-screw extruders in the general advantages they offer but, as has been mentioned, they are basically compounding and not extruding machines.

5.2.2.3 Internal Mixing

This is a batch process. Internal mixers of the Banbury type can be used for compounding reinforced thermoplastic materials. Their particular advantage is that the charge may be fully enclosed and confined under

TABLE 5.3

Compounding of Reinforced Thermoplastics in Single-Screw and Twin-Screw Extruders: Comparison of Some Important Process and Equipment Features

Single-screw	Twin-screw
No complete proprietary system available (for the manufacture of reinforced thermoplastics)	Systems offered
High wear	Low wear
Limited feed variants	A number of methods of introducing reinforcement
Difficulty in handling bulky feeds	Handles bulky feeds
Low capital cost	High capital cost
Short run versatility	Generally better suited to long runs
Ease of clean down	
Less complex to maintain	Maintenance more complex
Screw has no operational self-cleaning capability	Self-cleaning screw geometries available
Screw configuration fixed	Segmented screws and removable bushings facilitate screw modification in some systems

pressure so that incorporation of a bulky filler in the melt is facilitated. In modern commercial practice the equipment is not used for glass-reinforced thermoplastics: equally good mixing can be achieved in continuous processes with modern compounding extruders which offer greater facility and economy of production. Some use has been made of internal mixing in compounding thermoplastics (especially polypropylene) with asbestos fibre, but even with this comparatively bulky type of reinforcement continuous methods are now favoured.

Where an internal mixer is used the general compounding procedure is essentially similar to that which would be followed, say, in the manufacture of an unreinforced PVC compound. However, with a reinforcement to be incorporated two variants are possible. The polymer and reinforcement may be placed in the mixer at the same time and processed together from the start, or the polymer may first be fluxed in the mixer and the filler only then added. The shear effects will be greater in the former case, causing greater fibre breakdown if the filler is fibrous, and more heat will be generated. The wear of the working surfaces will also be greater and the ease of initial dispersion will be less. The reasons for this are analogous to those behind similar effects in twin-screw extrusion compounding when non-continuous fibre is being fed either jointly with the polymer into the conventionally placed feed throat, or separately into the molten polymer part-way down the barrel (*cf.* Section 5.2.2.1.2).

When both principal components of the charge (the polymer and reinforcing filler) have been added to the mixer they are contained under pressure exerted by a hydraulic or pneumatic ram, and compounded, by the action of the mixing rotors, at the temperature and for the time appropriate to the material. The compounded mix is discharged from a door in the bottom of the mixer on to a twin-roll mill or into an extruder, to be converted into sheet or strand which is subsequently cooled and granulated or pelletised.

5.2.3 Other Methods

The present chapter deals with methods of combining the base polymers with reinforced materials to form compositions which are then used, predominantly as moulding compounds, to manufacture finished products. The subject is thus manufacture of reinforced thermoplastics in the sense of compounding.

It is for this reason that methods of direct blending of the base polymer and reinforcement, which do not result in the formation of a solid integral compound (the 'reinforced thermoplastic material' in the context of the title of the present chapter), are mentioned in the present section, although they normally give loose mixtures converted directly into finished products by moulding. However they are dealt with at the end of the section after those miscellaneous, less commercially important methods which do lead to integral polymer/reinforcement compounds.

5.2.3.1 Miscellaneous Minor Methods of Production of Solid Polymer/Reinforcement Compounds

Many methods and method variants falling into this category have been put forward, mainly in patent specifications, but they have not found wide commercial application. Several such methods are mentioned in Appendix 1 in summaries of specifications of what might be termed 'process patents'.

Two methods are singled out for a mention below; one (the Rexall method) because of its technical interest in the historical context, and the other as an illustration of a non-conventional idea for the production of a fibre-reinforced thermoplastic material which can be prepared either as a moulding compound or directly as a finished product (sheet).

5.2.3.1.1 COAXIAL LIQUID COATING METHOD

This process is disclosed by the Rexall Drug & Chemical Co. in their British Patent 1 167 849 (claiming priority of the United States Application, October 1967). The technique involved is essentially one of liquid-coating continuous strands of glass fibres. The liquid is either a solution or an emulsion of the thermoplastic base polymer of the compound ultimately produced. Polystyrene and HD polyethylene emulsions, as well as polystyrene solution (in toluene), are mentioned in the examples as the coating media. The resulting product may be regarded as a non-extrusion analogue of the Fiberfil coaxial melt coating, in that after the glass roving has been passed through a bath of the polymeric coating medium the coating is dried and the polymer fused to give, after chopping into short lengths, parallel-fibre granules. Indeed one of the examples of the patent refers to impregnating glass rovings directly with 'a polypropylene resin melted in an extruder'. This would obviously be less efficient than the direct extrusion-coating with a melt as in the Fiberfil technique and therefore the Rexall method may be regarded as effectively limited to those polymers which can be applied in the form of emulsion, suspension or conveniently-handled solution.

5.2.3.1.2 THE FIBRE FUSION METHOD

The method, disclosed in ICI British Patent 1 200 342, essentially consists in making an intimate physical mixture in sheet form of two fibrous materials, one at least of which is a thermoplastic fibre, and fusing the thermoplastic fibre component (or the lowest melting component if more than one kind of thermoplastic fibre is present) to form a solid reinforced composition. This may be in the form of a sheet or may be directly moulded (e.g. by a compression method) into a form for which the initial fibrous mixture is suitable. The fused material may also be granulated and then used as a moulding compound. Mixtures of polypropylene and polyethylene terephthalate fibres, nylon and amosite asbestos fibres, polyethylene terephthalate and carbon fibres, and nylon and glass fibres are mentioned in the examples as precursors of the heat-consolidated products.

5.2.3.2 Polymer/Reinforcement Blends

These are essentially mixtures which are not converted into solid moulding compounds but are used directly in a fabrication process (moulding).

With this type of mixture the uniformity of ultimate reinforcement distribution is a function of the initial 'dry' blending and the further mixing action which occurs in the cylinder of the injection-moulding machine. The final uniformity of distribution of the reinforcement in the moulding may be less than in the case of pre-compounded stock, but the fibre breakdown can also be less because the mixture experiences shear in the hot polymer melt only once (during plasticisation and moulding—*cf.* also Section 5.3). Where direct blending is practised it tends to be more popular with asbestos-filled than glass-filled compositions.

The reinforcement and base polymer, the latter normally in powder form, may be pre-mixed in a blender and then fed to the injection-moulding machine or they may be metered directly into the hopper. The second method was the basis of the Dow 'black box' process[3,4] in which glass rovings are chopped above the hopper of an injection-moulding machine and metered-in alongside the polymer. Not all thermoplastic polymers are suitable for this process: polystyrene, polyolefins and SAN have been used.

Pre-blending in the dry form may be carried out in any convenient form of blender. Drum tumbling, ribbon mixers, the Henschel and Acrison mixers have been used, in particular for pre-mixing asbestos fibres with the base polymer (usually polypropylene). With this type of mix, stabilisers for the polymer which are normally used, are added at this stage.

5.2.4 Main Economic Considerations

A moulder will normally consider two main possibilities in deciding on the form and source of his moulding material.

(i) On-the-spot blending of polymer fibre and reinforcement (*cf.* Section 5.2.3.2).

(ii) Purchase of pre-compounded pellet stock (*cf.* Sections 5.2.1 and 5.2.2).

A variant of this is the use of pre-compounded concentrate stock, for dilution, as required, with unfilled material at the injection machine. Such concentrates are simply compounds with high reinforcement loading, and are marketed by many suppliers of pre-compounded materials.

In method (i) the material costs are comparatively low, and the physical properties of the mouldings may be somewhat better, because the material will not have undergone the mechanical working at elevated temperatures entailed by the compounding operation (*cf.* Section 5.3). However, the moulder alone is responsible for product quality (in method (ii) at least part of the responsibility rests on the compounder). Furthermore costs arise in connection with additional equipment and operations (blending, metering, material handling and conveyance). These costs may be quite high. Extra storage space may also be required, as may special equipment, e.g. an exhaust system when handling asbestos fibre. In some situations the additional expenditure may be balanced by the material cost saving only if the general scale of operation is sufficiently large.

The material cost in method (ii) is much higher but the need for auxiliary equipment is minimised or eliminated. In a moderate-size operation the masterbatch method variant may be cheaper (largely because the material cost works out somewhat lower) than the use of standard pre-compounded stock.

These points indicate the main general economic considerations, but the balance of all the factors, including technical, operational and product-quality aspects, must be carefully weighed up in each individual situation.

5.3 EFFECT OF COMPOUNDING ON PROPERTIES

Where a compounding operation, rather than simple 'dry' blending, is carried out to produce reinforced thermoplastic material (pellets or granules) for moulding, the properties of the material can be affected by the compounding process used, and—for a given process—by the severity of conditions within the limits characteristic of that process.

When the compound is moulded this second working at elevated temperature can further affect the properties of the composite material of the resulting moulding. The reasons for, and the nature of, the modifications are generally similar at both stages, although the effects may differ in magnitude.

The two main general sources of changes which processing can bring about in reinforced thermoplastic materials are modification of the polymer and modification of the reinforcing filler. In good processing practice polymer degradation is not normally a serious factor, although in systems (such as for example asbestos-reinforced polypropylene) where the polymer may be prone to thermal decomposition, with the tendency possibly aggravated by the presence of the filler, stabilisation and close attention to processing conditions may be necessary.

Fibrous fillers, most commonly glass or asbestos fibres, form the reinforcement in the great majority of reinforced thermoplastics. The fibres can and do break down under the shear experienced in compounding and even in moulding. This reduces the ultimate fibre length and affects the length distribution. Both factors, and in particular the first, are important parameters in the quality and degree of reinforcement provided by the fibres (*cf.* Chapter 1, Section 1.4).

In practice a certain amount of fibre breakdown is sometimes accepted or aimed for in compounding (see Section 5.2.2.1.1) to improve the surface finish of mouldings or as an effect attendant on high screw speeds for high output rates in single-screw extruders.

Studies of the effect of processing on the fibre parameters and hence on the material properties in relation to the injection-moulding process have shown that cylinder temperature, screw speed and back pressure are the main relevant factors.[5–7] This work is discussed further in Chapter 7.

Analogous effects in compounding have also received attention. Bernardo[8] demonstrated that:

(a) high shear energy input in melt-compounding HD polyethylene with glass fibres reduced the average fibre length to a value about three times lower than that for a sample compounded under low shear. The corresponding strength tensile and flexural and modulus values differed by a factor of about two in favour of the low-shear composite;

(b) the fibre length of two separate batches of chopped glass strand underwent the following changes after melt-compounding into HD polyethylene under standard conditions, followed by moulding on a screw-type injection machine:

Initial fibre length (predetermined by cutting)	0·250 in	0·125 in
Average fibre length after compounding	0·030 in	0·025 in
Average fibre length after moulding the compound	0·020 in	0·020 in

(c) the mechanical properties of mouldings produced directly from a 'dry' blend of the polymer and glass fibre were better than those of mouldings produced from pre-compounded pellets (pelleted concentrate, diluted as necessary with unfilled polymer). Some of these results are shown in Fig. 5.12.

Some effects of the blending or compounding method on the properties of polypropylene reinforced, respectively, with asbestos and glass fibres are shown in Tables 5.4 and 5.5.

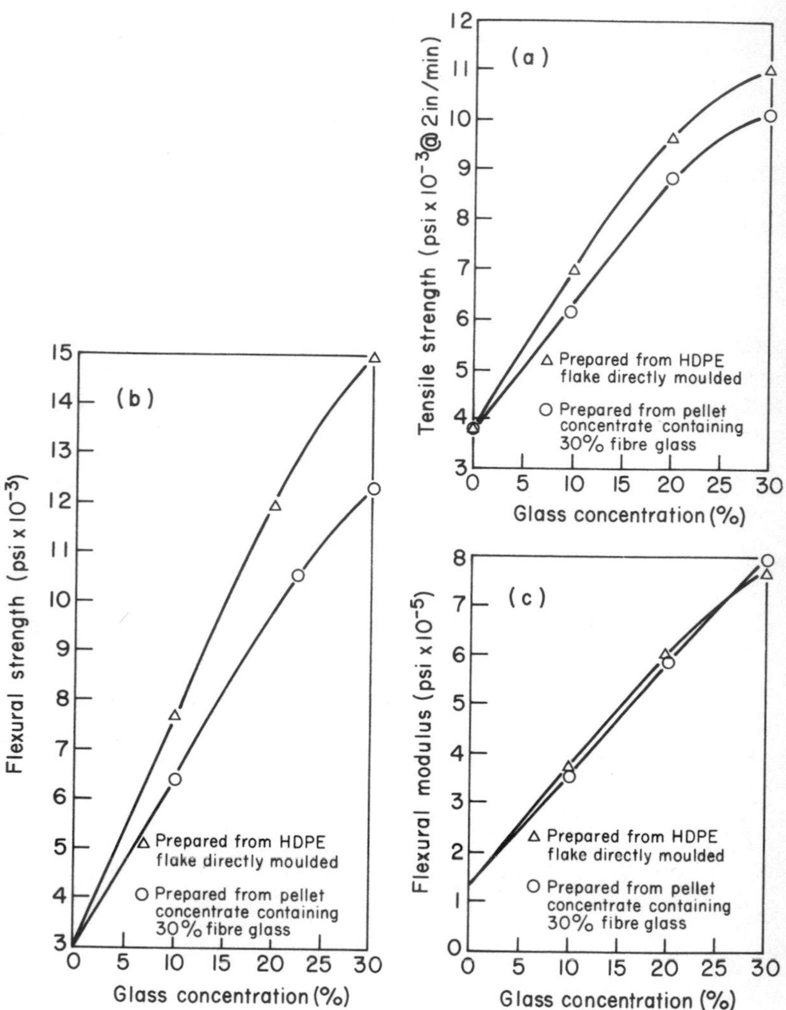

FIG. 5.12 Relationship of physical concentration of glass-fibre reinforcement of directly moulded flake and of prepared pellet in HDPE. (a) Tensile strength; (b) flexural strength; (c) flexural modulus. (Data of Bernardo: reprinted, with permission, from *SPE Journal*, **26,** October 1970.)

TABLE 5.4

The Effects of Mixing on the Physical Properties of Asbestos-Filled Polypropylene[a]
(Reproduced, with permission, from the Technical Bulletin 'Reinforced Polypropylene' (PLA-69-1622) of the Exxon Chemical Company, USA)

Property	Blending technique			
	Acrison blender	Drum tumbling	Henschel mixer	Banbury mixer
Tensile at yield (psi)	4 790	5 760	5 800	5 850
Elongation at yield (%)	8·4	5·8	6·8	7·5
Elongation at failure (%)	10·2	6·9	7·9	10·1
Izod impact (ft lb in^{-1})				
notched 72°F	0·56	0·46	0·50	0·46
unnotched 0°F	3·5	3·4	3·9	3·9
72°F	5·9	4·9	5·6	6·0
Flex secant mod. (psi)	299 000	309 000	314 000	318 000
Deflection temp. (°F)				
66 psi	—	284·5	—	—
264 psi	177	178	180	180
Melt flow rate (dg min^{-1})	0	0	0·38	0·53
Density (g cm^{-3})	1·05	1·04	1·03	1·03

[a] 80% powder, 20% chrysotile 7D3 asbestos.

TABLE 5.5

Comparison of Methods of Preparing Polypropylene/Glass Fibre Moulding Compounds[a]
(Reproduced, with permission, from the Technical Bulletin 'Reinforced Polypropylene' (PLA-69-1622) of the Exxon Chemical Company, USA)

Property	Method			
			C[d]	
	A[b]	B[c]	Short fibre	Long fibre
Secant flexural modulus (psi)	570 000	500 000	—	—
Izod impact strength (ft lb in^{-1})				
notched 72°F	4·0	1·7	—	—
unnotched 0°F	10·0	5·0	—	—
Estimated materials cost				
¢ per lb	24	31	56	77
¢ per 1 000 cm^3	55	72	131	178
Uniformity of dispersion	Best	Least	Intermediate	

[a] Basis: 20 wt % glass fibres.
[b] Resin powder dry-mixed with fibres by any of several methods.
[c] Mechanical mix of polypropylene pellets and reinforced-polypropylene pellets.
[d] Fully precompounded pellets.

REFERENCES

1. Fritch, R. and Fahr, G. (1959). *Kunststoffe,* **49,** 10, 543.
2. Werner & Pfleiderer Engineers, Stuttgart (FRG), KS—Information Brief Report No. 8, Continuous Production of Glass-fibre-reinforced Thermoplastics.
3. British Patent 1 055 395 (Dow Chemical Company), published 18th January 1967.
4. Jacopin, S. (1972). 27th Conference, RP/C Institute, SPI, Paper 11-C.
5. Filbert, W. C. (1969). *SPE Journal,* **25,** January, 65.
6. Lucius, W. (1973). *Kunststoffe,* **63,** 6, 367.
7. Maaghul, J. (1971). 26th Conference, RP/C, Division, SPI, Paper 8-A.
8. Bernardo, A. C. (1970). *SPE Journal,* **26,** 10, 39.

CHAPTER 6

PROPERTIES OF REINFORCED THERMOPLASTICS

6.1 GENERAL EFFECTS OF REINFORCEMENT

The main theoretical considerations relating to the effects of reinforcing fillers on the mechanical and physical properties of thermoplastic polymers have been outlined in Chapter 1 (Section 1.4), and some of these effects are illustrated by the data cited in Chapters 1, 3 and 4.

In this chapter the properties influenced or conferred by reinforcing fillers are reviewed in more detail: the tests commonly employed to determine these properties are discussed, with reference to the practical use made of the information they yield by the designer and user, and typical data are quoted for a range of properties in a number of thermoplastic polymers: the properties are also compared with those of some thermosetting polymers and metals, with which reinforced thermoplastics compete especially in engineering applications.

As has been mentioned in Chapter 4, it is the mechanical as well as some physical properties of the base polymers which, by definition, benefit to the greatest extent, both in the short- and the long-term aspects, by the incorporation of reinforcing fillers. The temperature effects on the short-term and long-term mechanical properties are also modified, very strongly in some cases, as are environmental effects (environmental stress cracking, weathering). In general the incorporation of the reinforcing fillers with the greatest all-round effect (e.g. glass fibres, carbon fibres) in thermoplastics increases the short-term stiffness and strength, increases creep resistance and fatigue endurance, and reduces the effect of temperature on both these groups of properties. Dimensional stability is also improved and thermal expansion as well as mould shrinkage reduced.

Impact strength (essentially a matrix-associated property, *cf.* Chapter 1, Section 1.4), which can be related to the toughness of the material, is reduced in some thermoplastic composites. The effect is influenced by several factors. Thus filler loading, especially to a high level, may alter the mode of failure from ductile to brittle: this would reduce the impact strength, and also tend to lower the fatigue strength, of the composite.[1] The reduction may, however, be counteracted, at least to some extent, if the interfacial adhesion between the polymer matrix and the reinforcement is of the right magnitude (i.e. not too high) to permit energy absorption on impact by de-bonding,[2,3] in a way similar to that in which impact energy is absorbed by fibre-reinforced thermosetting resin composites. At low reinforcement loading, or generally in systems where matrix-governed ductile failure occurs, increased interfacial adhesion (e.g. through coupling) between reinforcement fibre and matrix can increase impact strength.[2] Part of the reinforcing action of fibrous fillers is the restriction of crack propagation in the matrix. The effectiveness of this mechanism increases with fibre length; hence the latter is also a factor in the effect of reinforcing fillers on the impact strength of reinforced thermoplastics.[4]

6.2 PROPERTIES OF PRACTICAL INTEREST AND THEIR DETERMINATION

This section is concerned with those properties of reinforced thermoplastics, as well as the tests commonly used for their evaluation, which are of direct practical interest to the producer and user of these materials. The theory of plastics testing and test design is neither directly relevant here, nor within the general plan and scope of this book. The reader interested in the fundamentals and broad theory of the subject *per se* may care to refer to two informative books, respectively by Turner[5] and a group of ICI scientists.[6]

In practical terms, the object of a test to determine a property of a material is to obtain a numerical value which is sufficiently accurately descriptive of the property to be useful for the purpose of comparison with other materials, or the prediction of performance in service situations. Both elements are involved in material selection and product design. Property comparison in terms of numerical results of tests is also the basis of quality control.

The main 'external' factors which influence the properties, and in particular the mechanical properties, of a plastics material are the level of stress to which the material is subjected, the time over which the stress operates, the temperature, and the effects of the environment. This applies to any material, although the scale or range of each factor for a particular level of effect can be different for different materials. With materials used for engineering purposes the effect of all the factors, within limits appropriate to the end-use conditions, must therefore be considered, with special reference to the maximum deformation (strain) that may be tolerated. Because of their chemical structure and morphology, thermoplastics are affected by all the above factors within the moderate ranges and scales of magnitude associated with normal service conditions. Incorporation of reinforcing fillers increases the strength and reduces the effects of time under stress and of temperature on the mechanical properties of thermoplastics—the group of properties most important in their engineering applications. The latter may be broadly defined as those applications in which the part or moulding is required to sustain a load either continuously or at intervals during its service life, and/or retain its form, stiffness and strength at elevated temperatures in the way in which they would be retained by a metal or ceramic part.

Tests which evaluate the effect of time upon the property being measured are sometimes styled 'long-term' tests, the properties so determined 'long-term' properties and the data obtained 'multiple-point' data, in contradistinction to the 'single-point' data obtained in short-term tests. Multiple-point data are especially relevant to the prediction of the performance of a plastics material in service.

Thought is increasingly being given to rationalising the consideration and standardising the presentation of plastics design data,[7,8] with a view, *inter alia,* to helping the designer in making a rational selection of a material. The relevant systematically assembled data are sometimes collectively described as 'the field of information' (or 'spectrum of information') and it is possible through suitable analysis to arrive at the design of a tool which may (perhaps after slight modification following initial trials) serve as a production tool. However, the more direct practical approach is probably still the most common. Thus the initial material selection will be made in the light of direct experiencc of the performance and properties of candidate materials in similar applications or, if the experience is limited or lacking, on the basis of a comparison of available 'single-point' short-term test data. In the former case a tool may be produced and final adjustments madc after some prototype parts have been moulded. In the second case a prototype tool is invariably first produced and then modified as necessary.

6.2.1 Mechanical Properties

6.2.1.1 Short-term Mechanical Properties and Tests

These properties are determined in tests in which a standard specimen is stressed to the point of failure (or specified degree of deformation) under strictly prescribed conditions. The value of the breaking or deforming stress (or load) is taken as a measure of the property. Strict specification of the conditions and the form of test specimen is necessary, because differences, e.g. in magnitude or rate of application of stress, or in specimen dimensions, will result in different numerical values. The American ASTM standard tests are still the most widely used with reinforced thermoplastics.

It may be noted that, as pointed out by Turner,[9] the *physical nature* of the failure phenomena is normally the same in both the short- and long-term tests; the main difference lies in the significance to the result of the particular phases of the failure process.

The main, proper use of short-term test data is for initial comparison of materials, preliminary selection and quality control, and as an indication of the effect of processing conditions on a material (the same or similar grades).

In evaluating the effects of reinforcement in thermoplastics (which is essentially comparison of materials) the data from short-term determinations on the base polymer alone may be compared directly with corresponding test results for the polymer/reinforcement composite, as has been done, e.g. in the tables of Section 3.3 (Chapter 3).

The comparison may also be made in terms of ratios of the values of the same property of the composite and the unreinforced base polymer. Such ratios of the flexural modulus values and strength (impact or tensile) values respectively have been called the 'stiffness factor' and 'reinforcement factor.'[2,3] This method of comparison is convenient for some purposes: for example when the effect of reinforcement is to be illustrated in relation to some variable (e.g. reinforcement content, temperature, strain) in one and the same polymer or the effects in a number of polymers are compared on this basis. This relative basis of presentation also provides a good way of illustrating the effect of orientation and length of reinforcement fibres upon the property considered, as exemplified by Fig. 6.1.

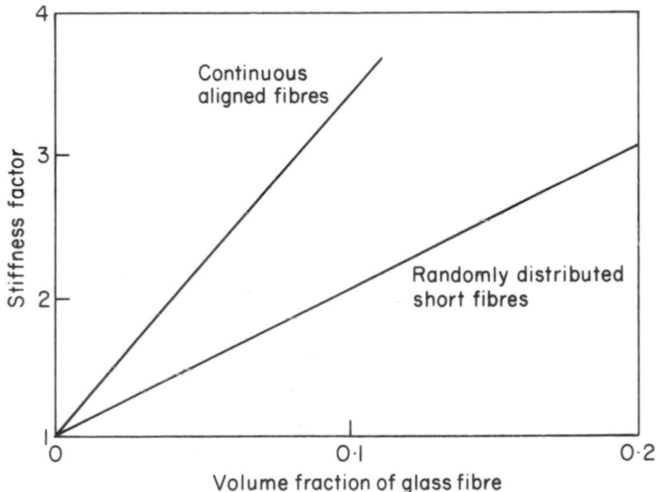

FIG. 6.1 Stiffness factor versus volume fraction of glass fibres for a reinforced nylon with the fibres aligned and randomly distributed. (Data of Ogorkiewicz:[2] reproduced by permission of *Composites*, IPC Science and Technology Press Ltd.)

Comparison of appropriate stiffness factor plots provides a particularly good method of demonstrating and assessing the effects of interfacial adhesion (coupling) between reinforcing fibres and polymer matrices.[5]

6.2.1.1.1 STANDARD DEFINITIONS AND TEST METHODS (MECHANICAL AND SOME PHYSICAL PROPERTIES)

In the list a brief definition of the property (mainly based on ASTM standards) is followed by the number of the ASTM standard relevant to reinforced thermoplastics, with equivalent international (ISO), German (DIN), or British (BS) standards in the same line. Standards, and any directly relevant non-standard methods, dealing with the same property but not fully equivalent (e.g. employing different test methods, specimens and/or conditions) are entered vertically below the ASTM reference. Units in which the property value is expressed follow in the order: Imperial, metric, SI. The relevant conversion factors are given in Appendix 3.

A comprehensive review of standard test methods for plastics materials has been published by Ives, Mead and Riley.[10]

(i) *Tensile strength*

At yield: maximum tensile stress sustained by the specimen at the yield point in a tensile test.

At break: maximum tensile stress sustained by the specimen if it occurs at break.

Tests: ASTM D-638*; ISO R 527; DIN 53 455
ASTM D-759†
ASTM D-1708‡

Units: psi; kgf cm^{-2}; N m^{-2}

(ii) *Percentage elongation*

The elongation of a test specimen (at yield or break) expressed as a percentage of the original ('gauge') length

Tests: as under (i)

Units: %

(iii) *Tensile modulus* (*Young's modulus; modulus of elasticity*)

The ratio of (nominal) tensile stress to corresponding strain below the proportional limit of the material.

Tensile stress (*nominal*): the tensile load per unit area of minimum original cross-section within the gauge boundaries, carried by the test specimen at any given moment.

* Determination at room temperature (73°F).
† Determination at −40°F or 170°F
‡ Microtensile specimens.

Proportional limit: the greatest stress which a material is capable of sustaining without any deviation from proportionality of stress to strain (Hooke's Law).

Tests: as under (i)

Units: as under (i)

(iv) *Flexural strength*

At break: maximum stress in the outer fibre of the specimen at the moment of break in bending.

At yield: stress at yield (calculated from formulae appropriate to the test method).

Tests: ASTM D-790; ISO R 178; BS 2782
DIN 53 452

Units: as under (i)

(v) *Flexural modulus*

Tangent modulus of elasticity: the ratio, within the elastic limit, of stress to corresponding strain in bending, calculated by the appropriate theory.

Secant modulus of elasticity: the ratio of stress to strain at any given point on the stress–strain curve.

Tests: as under (iv)

Units: as under (i)

(vi) *Modulus of rigidity* (*shear modulus*)

The ratio of (nominal) shear stress to corresponding strain, below the proportional limit of the material. Usually determined by testing in torsion.[5]

In the method of ASTM D-1043 the value obtained by measuring the angular deflection resulting when a specimen is subjected to an applied torque is referred to as 'apparent' modulus of rigidity.

For materials which conform to the laws of classical elasticity, the modulus of elasticity (*cf.* (iii) above) is related to the shear and bulk moduli respectively by the formulae:

$$E = 2G(1+\lambda)$$
$$E = 3K(1-2\lambda)$$

where: E = modulus of clasticity, G = modulus of rigidity, K = bulk modulus, λ = Poisson's ratio (ratio of change in width per unit width to change in length per unit length).

Tests: ASTM D-1043; ISO R 537
ASTM D-2236

Units: as under (i)

(vii) *Shear strength*

The maximum load required to shear the specimen so that the moving portion has completely cleared the stationary portion.

Tests: ASTM D-732
BS 2782
Units: as under (i) N.B. values based on the area of sheared edge or edges.

(viii) *Compressive strength* (*Crushing strength*)
The maximum compressive stress (nominal) carried by a test specimen during a compression test (this may or may not be at the moment of rupture)—N.B. Compressive yield strength: normally the stress at yield point in compression.
Tests: ASTM D-695
DIN 53 454
Units: as under (i)

(ix) *Impact strength*
In pendulum-impact methods: energy expended by a standard pendulum-impact machine to break a standard test specimen (plain or notched) under stipulated conditions of specimen mounting, notching (stress-concentration) and pendulum velocity at impact.
In falling-weight methods: the minimum value of the product of fall height and mass to cause fracture of a standard test specimen.
There are two pendulum methods in common use:
(a) *Izod Method*: specimen held (clamped) as a cantilever beam is broken by a blow delivered at a fixed distance from the edge of the clamp. The standard ASTM method requires a *notched* specimen: the notch is intended to produce a standard stress concentration.
Tests: ASTM D-256 (Method A or C)—at room temperature (73°F)
ASTM D-758 at −40°F
ISO R 180 at 20°C
BS 2782 (Method 306A): this is similar to D-256 but uses a sharper notch, giving lower relative values
**Units*: ft lb per in of notch
cm kgf per cm of notch
N m per m of notch (N.B. N m = J)
(b) *Charpy Method*: the specimen, which may be plain or notched, is supported as a simple beam (at two ends) and broken by a blow delivered midway between the supports.
Tests: ASTM D-256 (Method B)
ISO R 179; DIN 53 453
BS 2782 (Method 306 E)
**Units*: ft lb per in of notch
cm kgf per cm of notch
N m per m of notch

* Some standards still use energy per unit *length* of notch but energy per unit *area* (i.e. in^2, cm^2, m^2) of the cross section of the unnotched part of the sample is also often used.

The falling-weight method is represented by that of BS 2782 (Methods 306B and C) in which a ball of known weight is dropped centrally onto a disc specimen supported annularly by a hollow cylinder.

The tensile impact method measures the energy required to rupture by shock in tension a standard specimen in a pendulum-impact machine in which one end of the specimen is mounted in the pendulum and the other is gripped by a special mounting, which initially travels with the pendulum but is subsequently arrested at the instant of maximum pendulum kinetic energy.

Test: ASTM D-1822

Units: ft lb; kgf cm; N m

The impact strength as measured in the above tests is a complex property,[10] sensitive to various factors, including rate of straining,[19] notch type and others, and the tests used are arbitrary. Nonetheless the test results, especially with unnotched specimens, are a useful practical indication of the toughness of reinforced thermoplastics. Theberge and Hall[19] found a high degree of correlation between *unnotched* Izod impact strengths and tensile impact strengths of several glass-fibre-reinforced thermoplastics, but notched Izod impact strengths did not correlate with either. With some materials, e.g. structural foam (*cf.* Chapter 10) notching may also disturb the structure (surface skin) and notched specimens are therefore inappropriate.

(x) *Hardness*

This is normally measured in terms of resistance to indentation under rigidly specified conditions (with special reference to indentor type and geometry, load applied and method of loading).

Tests: Numerous: the Rockwell test is often applied to reinforced thermoplastics:

ASTM D-785 (Rockwell hardness)
DIN 53 456 (Ball indentation hardness)
ISO DR 2039 (Ball indentation hardness)

Units: Arbitrary; specific to method. Include method designation and scale symbol (e.g. Rockwell M).

(xi) *Deformation under load*

Percentage dimensional change under compressive load in specified conditions.

Test: ASTM D-621

Units: %

(xii) *Mould Shrinkage*

Difference between a linear dimension of the specimen (after prescribed conditioning) and the corresponding dimension of the mould cavity, divided by the latter.

Test: ASTM D-955

Units: properly none, but value usually expressed in inches per inch or millimetres per millimetre.

(xiii) *Specific gravity*

The ratio of the weight in air of a unit volume of the material at 23°C to the weight (determined in identical conditions) of an equal volume of gas-free distilled water. For practical purposes numerically equal to the DENSITY (defined as the weight in air of a unit volume of material at 23°C).

Tests: ASTM D-792 ISO R 1183
DIN 53 479
BS 2782 (Method 509A)

Units: Specific gravity: none
Density: lb ft^{-3}; g cm^{-3}; kg m^{-3}

(xiv) *Water absorption*

Percentage increase in weight of a specimen after immersion in distilled water: various immersion periods (e.g. 2 h, 24 h, to saturation), may be used, at various temperatures (e.g. 23°C, 50°C, at boil).

Tests: ASTM D-570
ISO R 62; DIN 53 475
BS 2782 (Method 502)

Units: Appropriate to purpose and method: e.g. weight % or weight per time period.

The above standard tests are aimed at determining the water content reached after a given time at a given temperature. In many end-use situations (and for the purposes of study) it is important to know the *rate* of sorption (which determines the time to equilibrate), and the equilibrium content (which may not be reached in arbitrarily fixed test conditions if the rate of sorption is slow). The sorption rate is governed by the diffusion behaviour of the penetrant (here water) in the polymer, characterised by a physical parameter known as the diffusion coefficient or diffusivity, with the dimensions L^2T^{-1} and units of $cm^2\ s^{-1}$ or m^2s^{-1}. Diffusion coefficients vary with temperature in all polymer/water systems; in many they are also dependent on the concentration. Diffusion of water in polymers and the determination of diffusion coefficients in such systems have been discussed, *inter alia,* by Barrie,[11] Braden[12] and Stafford and Braden.[13] Informative more general treatments of diffusion of penetrants in polymers can be found in publications by Fujita,[14] Crank,[15] and Meares.[16,17] Diffusion in heterogeneous polymer systems (filled polymers) was investigated by Barrer.[18]

6.2.1.1.2 SOME TYPICAL SHORT-TERM PROPERTY DATA FOR REINFORCED THERMOPLASTICS AND OTHER RELEVANT MATERIALS

Short-term values of properties important in engineering applications are given for several thermoplastics in Figs. 6.2–6.4 and compared with those for certain metals and thermosetting composites.

The values shown are either the highest reported for the particular commercial composite, or means of a range. For directionally reinforced materials (e.g. epoxy/carbon fibre composite) the data are for the direction of greatest reinforcement.

Other data have been given in Chapters 3 and 4.

6.2.1.1.3 EFFECTS OF THE LEVEL OF LOADING AND CHARACTERISTICS OF FIBROUS REINFORCEMENT IN REINFORCED THERMOPLASTICS

The ways in which the amount of fibrous reinforcement and the fibre characteristics can influence the mechanical properties of a reinforced thermoplastic were mentioned in Chapter 1, Section 1.4. Some actual effects of the level of loading on the strength and modulus of thermoplastics are shown in Figs. 6.5 (glass-fibre reinforcement) and 6.6 (asbestos-fibre reinforcement). As can be seen from the figures the tensile strength and stiffness generally increase with increasing amount of reinforcement up to quite high levels of loading, but the effect (especially on tensile strength) tends to be less with asbestos fibre.

In the design of parts for engineering applications the impact behaviour of plastics—which is related to their toughness—ranks in practical significance with the two most important long-term properties: creep and fatigue endurance (see Section 6.2.1.2). Unfortunately, because the results of impact strength determinations are strongly dependent on the test conditions and represent a rather complex property sensitive to several factors, they cannot be utilised directly in design formulae. However, when used with proper caution and interpreted in terms of relevant experience, they are a valid aid to material comparison and preliminary selection. The impact strength of reinforced thermoplastics, as measured in terms of the standard tests, increases with reinforcement content in some materials and decreases in others; i.e. on this basis some composites become tougher and some more brittle with increasing reinforcement loading. The effects are illustrated, for some commercial glass-filled thermoplastics, in Figs. 6.7 and 6.8. However, as has been mentioned (*cf.* Sections 6.1 and 6.2.1.1.1 (ix) above), correlations between the results of different tests can be poor. It has been suggested (by Theberge and Hall[19]) that those glass-fibre-reinforced thermoplastics which are regarded as the toughest in actual service are the ones which are the most sensitive to rate of straining in impact tests. Within limits, the initial glass fibre length is not regarded as an important factor in the ultimate impact strength,[19] but the effective fibre length has been claimed (by Wilson[20]) to

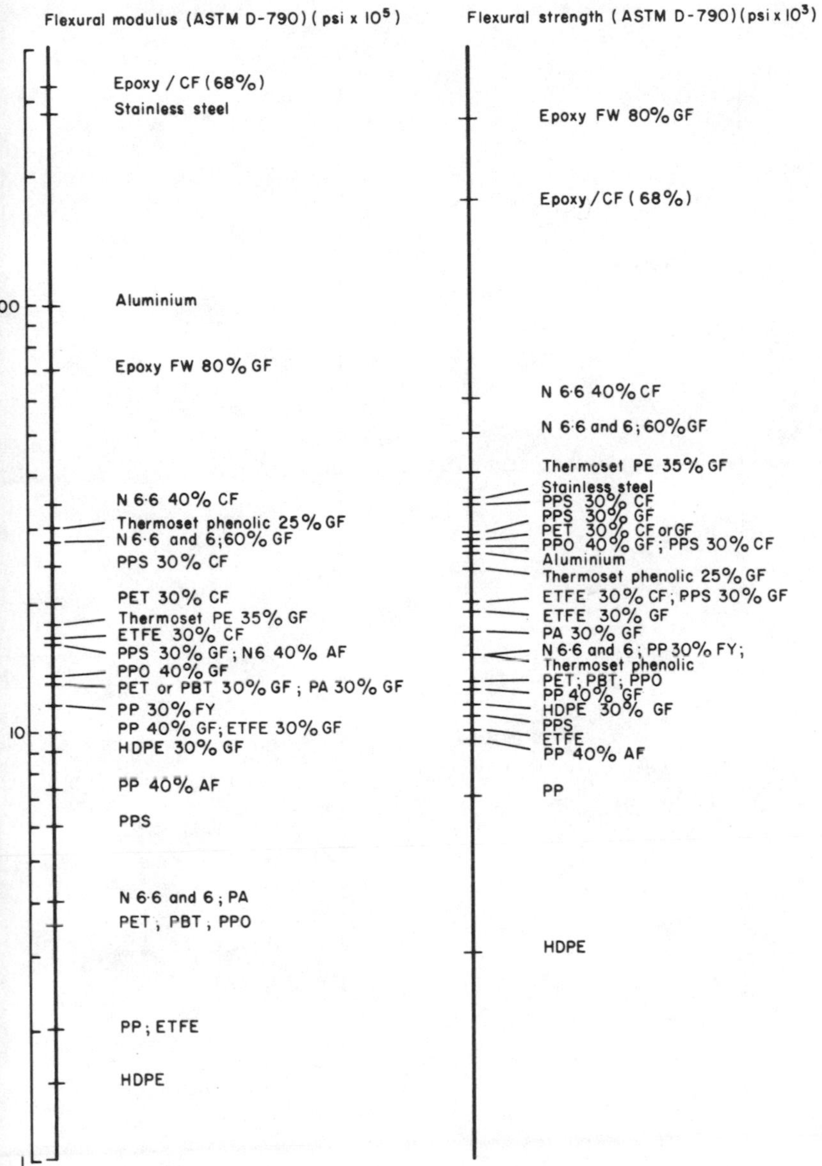

FIG. 6.2 Flexural modulus and strength values of some reinforced thermoplastics and other materials. FW = filament wound, other abbreviations as in Appendix 2. Logarithmic scale refers to both portions of the diagram, with the appropriate factor as shown above each axis.

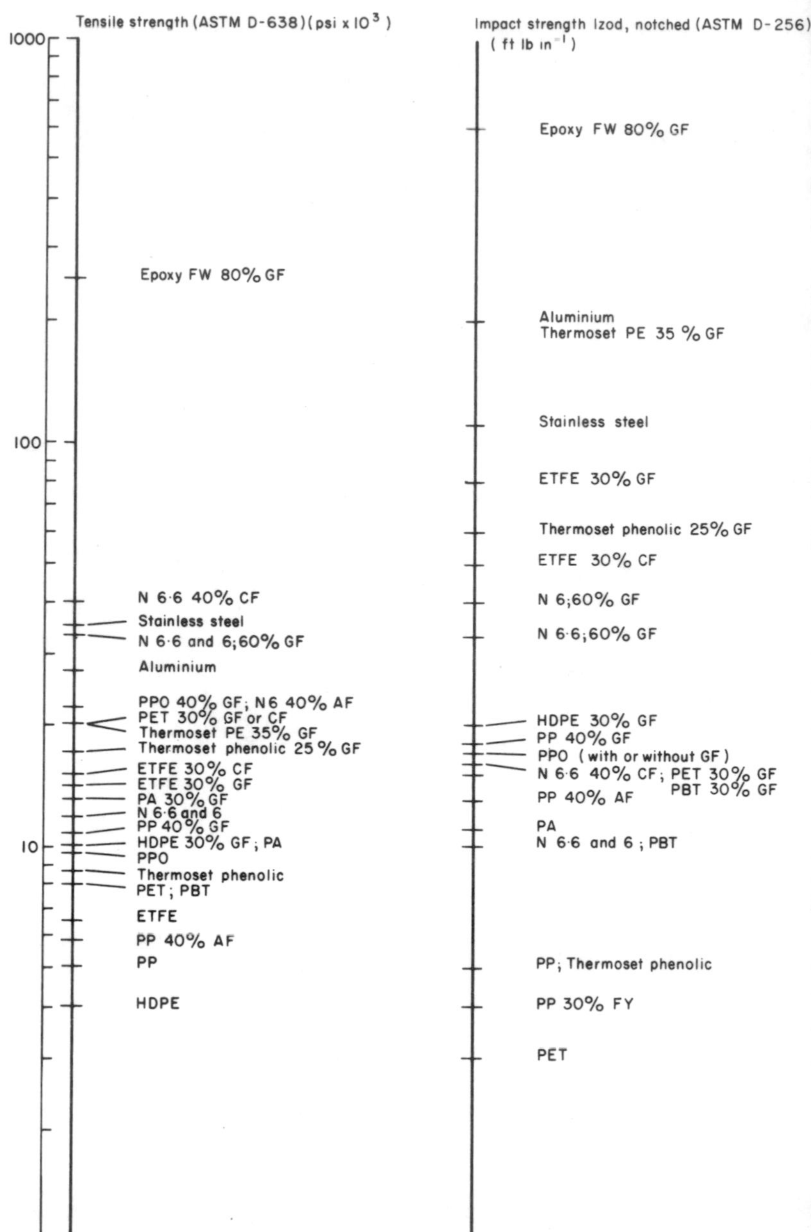

FIG. 6.3 Tensile and impact (notched Izod) strengths of some reinforced thermoplastics and other materials. Abbreviations as in Fig. 6.2. Logarithmic scale refers to both portions of the diagram, with the appropriate factor as shown above each axis.

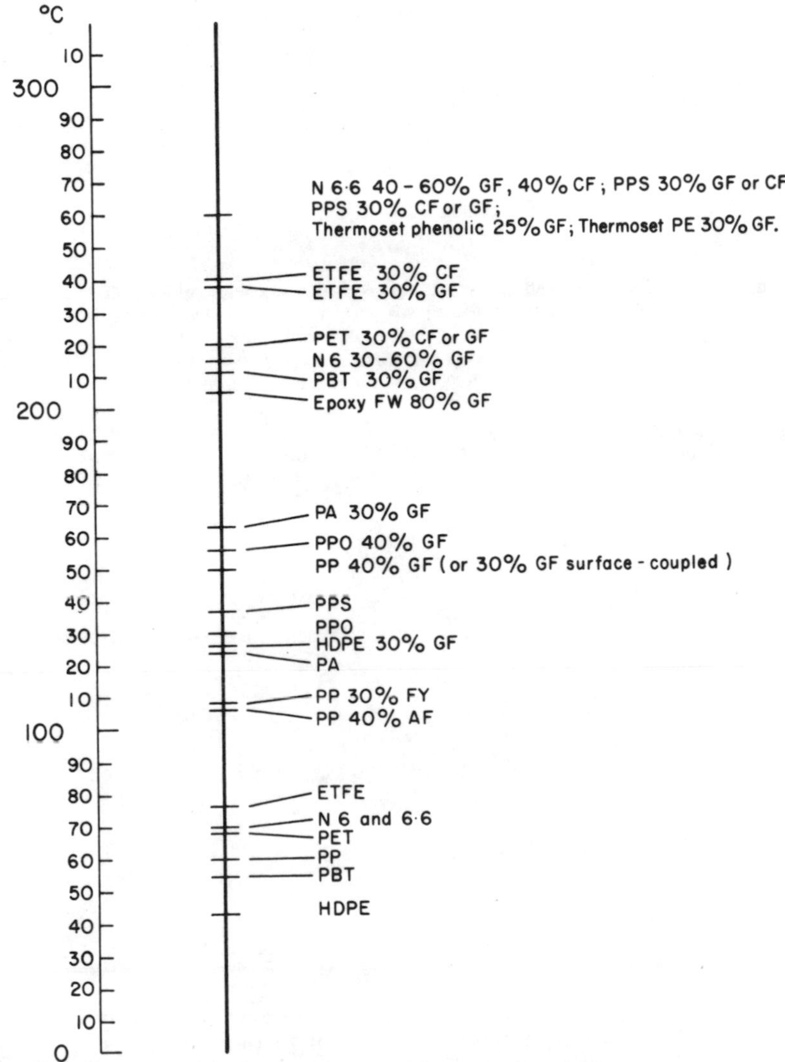

FIG. 6.4 Heat distortion temperatures under load (264 psi): reinforced thermoplastics and some thermoset composites. Abbreviations as in Fig. 6.2.

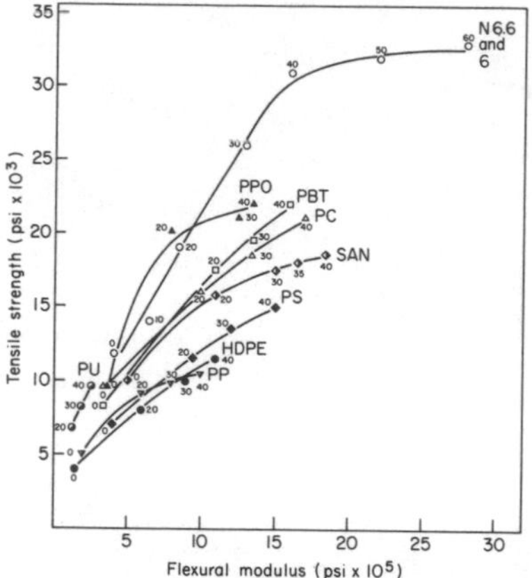

FIG. 6.5 Flexural modulus and tensile strength as functions of glass-fibre content of some reinforced thermoplastics. Points on the curves correspond to the wt % glass-fibre content given by the figures. Abbreviations as in Fig. 6.2. (Thermocomp series: Liquid Nitrogen Processing Corp.) *Note*: Figures for 'surface-coupled' glass-fibre-filled polypropylene (Hercules Pro-fax PCO72) show the degree of improvement:

Glass % by weight	10	20	30
Flexural modulus (psi)	$4{\cdot}4 \times 10^5$	$6{\cdot}3 \times 10^5$	$8{\cdot}5 \times 10^5$
Tensile strength (psi)	9×10^3	13×10^3	15×10^3

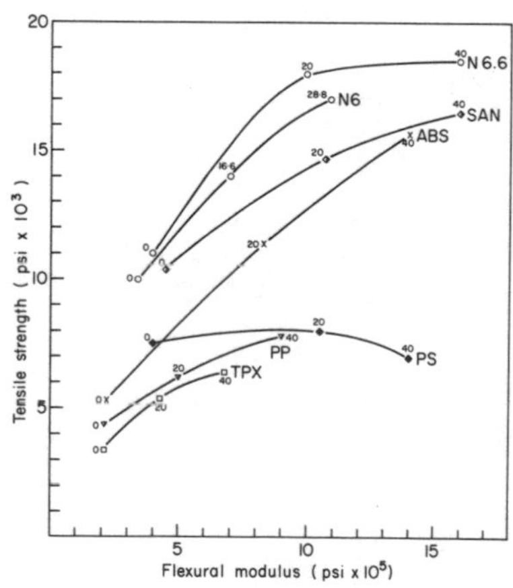

FIG. 6.6 Flexural modulus and tensile strength as functions of asbestos-fibre content of some reinforced thermoplastics. Points on the curves correspond to the wt % asbestos-fibre content given by the figures. Abbreviations as in Fig. 6.2. (Data of Hollingsworth,[22] Crown copyright, reproduced from *Composites* with the permission of the Controller, H.M. Stationery Office.)

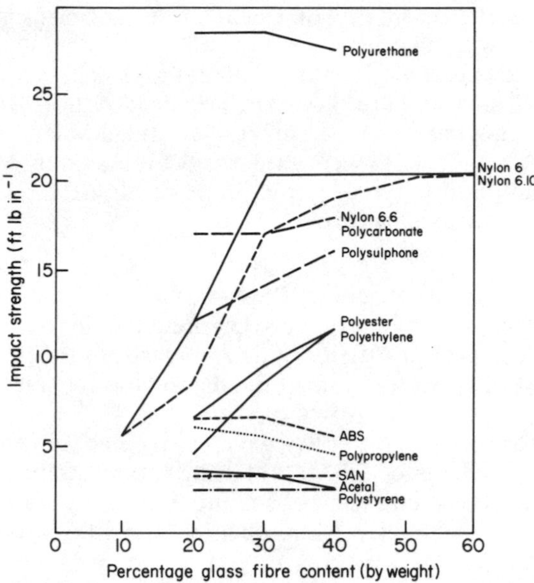

FIG. 6.7 Unnotched Izod impact strength *vs* glass-fibre content. (Lanham's data.[23])

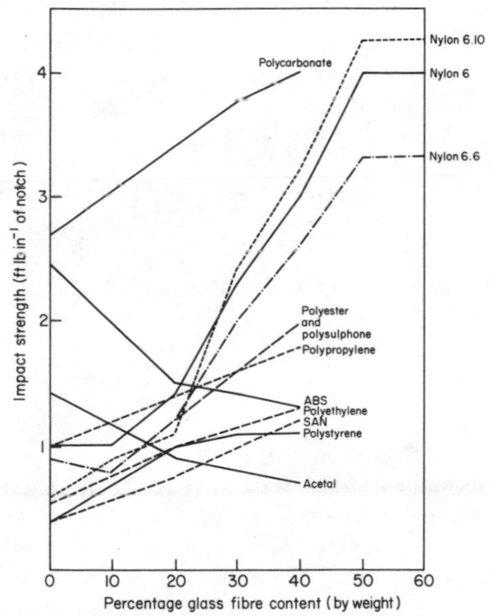

FIG. 6.8 Notched Izod impact strength *vs* glass-fibre content. (Lanham's data.[23])

influence the notch sensitivity of impact test specimens (significantly lower sensitivity with fibre length $\geq \frac{1}{4}$ in).

The combined effects of glass-fibre content and such characteristics of this reinforcing filler as fibre diameter, original length (chopped strand length), thickness of the strands (fibre bundles), and amount of size present on the fibres ('strand solids') were extensively evaluated by Maaghul[21] in glass-filled nylon and polypropylene. Some of his data relevant to the subject of the present section are shown in Figs. 6.9–6.12.

6.2.1.2 Long-term Mechanical Properties

From the practical standpoint of their determination in tests and effects in service the mechanical properties of thermoplastics may be regarded as time-dependent in two senses (albeit the differentiation is to some extent artificial).

Where increasing stress (in practice increasing load) is applied the rate of application, and the associated rate of deformation (strain) can influence the result of a test, or end-use behaviour, principally by changing the mode of failure, commonly from ductile to brittle with sufficiently increased straining rate.

Over periods of time at constant stress (in practice usually constant load) changes can occur in strain, and vice versa, giving rise, respectively, to the phenomena of creep and stress relaxation: strength can also decrease—to the point of failure—with time under load, continuous or intermittent (static or dynamic fatigue). It is this second group of time effects that constitutes the so-called 'long-term' mechanical properties, highly important in engineering design.

Time-dependence of the relationships between stress at constant deformation and deformation at constant stress is a cardinal feature of viscoelastic behaviour. In thermoplastics this behaviour is predominantly non-linear. It arises, in a complex manner, in consequence of the molecular structure and morphology of these materials. A full description of this aspect is beyond the scope of the present section: several works provide a useful introduction to the subject.[6,24–29]

6.2.1.2.1 CREEP

Nature, determination and presentation of data

Creep is the change of strain with elapsed time at constant stress (in practice constant load), when the temperature and any other conditions which may affect the relevant properties of the material (e.g. relative humidity) are kept constant. One of the relevant standards (ASTM D-674) defines creep as 'the time-dependent part of the strain which results from the application of a constant stress to a solid at a constant tempera-

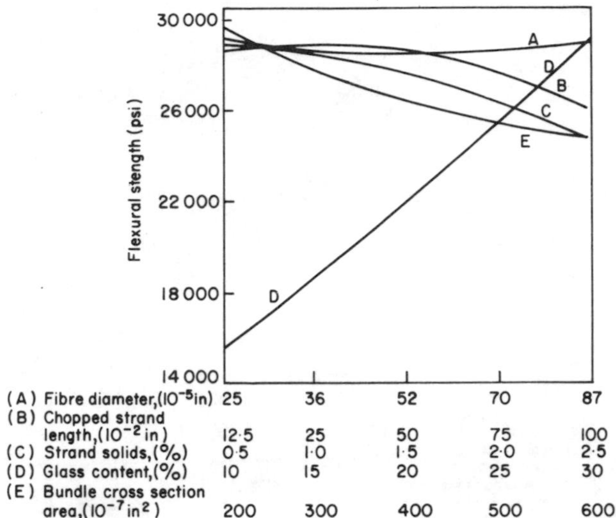

FIG. 6.9 Flexural strength property of nylon and its interdependence with fibre-glass construction variables. (Data of Maaghul: Proceedings of the 26th Annual Conference RP/C Division, SPI, 1971, Paper 8-A, reproduced with permission of the author.)

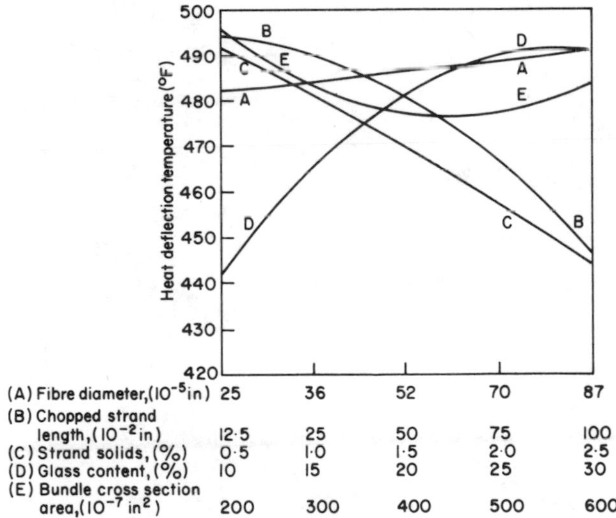

FIG. 6.10 Heat deflection temperature of nylon and its interdependence with fibre-glass construction variables. (Data of Maaghul: Proceedings of the 26th Annual Conference RP/C Division, SPI, 1971, Paper 8-A, reproduced with permission of the author.)

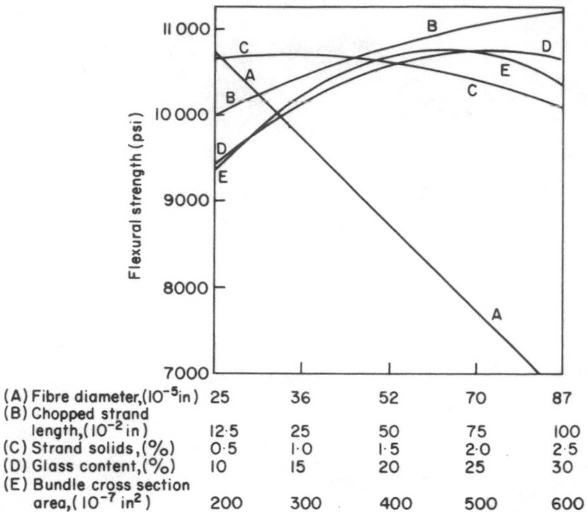

FIG. 6.11 Flexural strength property of polypropylene and its interdependence with fibre-glass construction variables. (Data of Maaghul: Proceedings of the 26th Annual Conference RP/C Division, SPI, 1971, Paper 8-A, reproduced with permission of the author.)

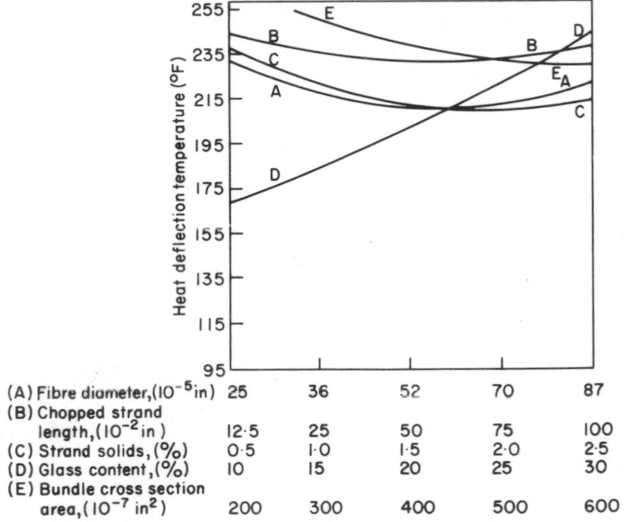

FIG. 6.12 Heat deflection temperature of polypropylene and its interdependence with fibre-glass construction variables. (Data of Maaghul: Proceedings of the 26th Annual Conference RP/C Division, SPI, 1971, Paper 8-A, reproduced with permission of the author.)

ture. That is the creep at a given elapsed time is equal to the total strain at the given time minus the instantaneous strain on loading'.

In a creep experiment the strain (change in relevant dimensions) is determined as a function of time under a load applied to the specimen at the outset to produce the required mode of stress (tension, shear, flexure, etc.), and remaining constant throughout. Several experiments are usually performed, with different constant stress values. The results are represented directly by plots of strain against time (usually log time)—Fig. 6.13 (i). Other curves, derived from this type of plot as shown in Fig. 6.13, are also frequently used to present creep data.

Creep compliance may be calculated from the ratio of strain to stress represented by any point on the curves of Fig. 6.13 (i).

The 'creep modulus' plotted in Fig. 6.13 (iv) is the reciprocal of creep compliance.

As shown schematically in Fig. 6.14, a creep experiment may be followed by determination of strain recovery by removing the stress (load) and

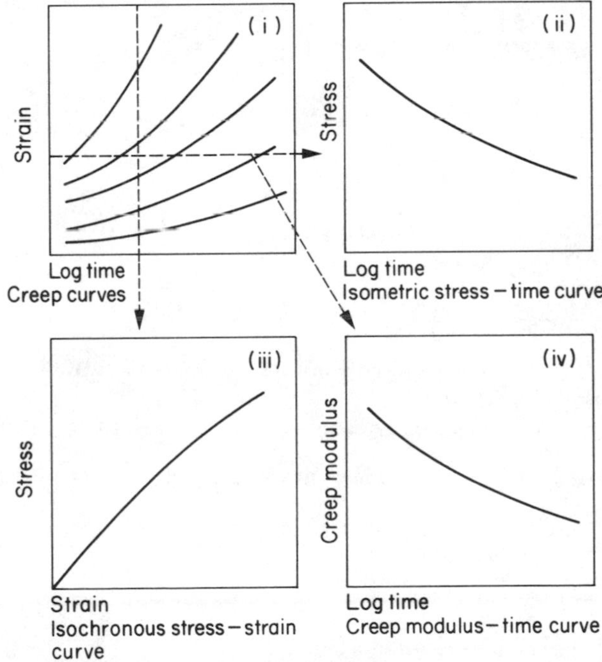

FIG. 6.13 Graphical representation of creep data. The relationship represented by graph (iii), i.e. the ratio of initial applied stress to the total strain at time t, is sometimes called the 'apparent modulus'. (Reproduced, with permission, from TS Note N110 (Third edition), ICI Plastics Division.)

measuring the decreasing strain as a function of time. The experiment is normally carried out at constant temperature, but recovery may be accelerated by the application of heat in an actual service situation. Creep recovery in thermoplastics can be complete unless an irreversible change has taken place (*cf.* Fig. 6.14).

The British Standard specification which deals (*inter alia*) with the significance of creep in plastics part design,[7] recommends the representation of strain recovery behaviour in the form of a plot of fractional recovered strain *vs.* reduced time. The general form of such a plot is shown in Fig. 6.15.

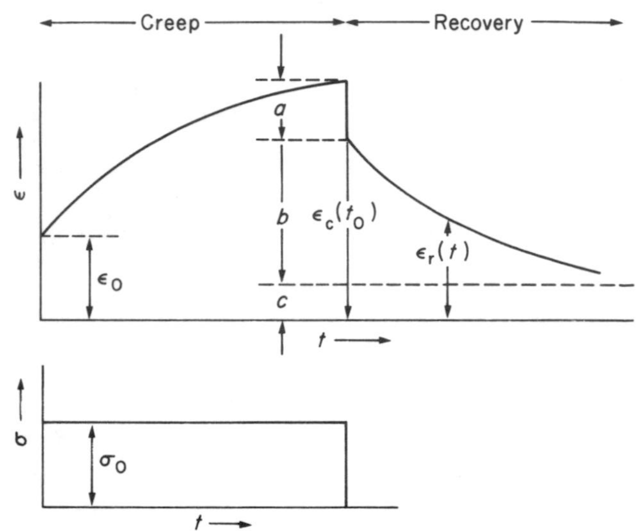

FIG. 6.14 Creep and strain recovery, schematic representation. (Partly after Joisten.[30])

ε = strain, δ = stress; t = time
a = instantaneous elastic recovery (may not occur in some thermoplastics)
b = delayed recovery
c = irreversible change if present (e.g. stress cracking, crystallisation, deformation past the yield point)
$\varepsilon_c(t_0)$ = creep strain at the time of removal of stress
$\varepsilon_r(t)$ = strain remaining at time t
t_0 = duration of creep period

$$\frac{\varepsilon_c(t_0) - \varepsilon_r(t)}{\varepsilon_c(t_0)} = \text{fractional recovered strain}$$

$$\frac{t - t_0}{t_0} = \text{reduced time}$$

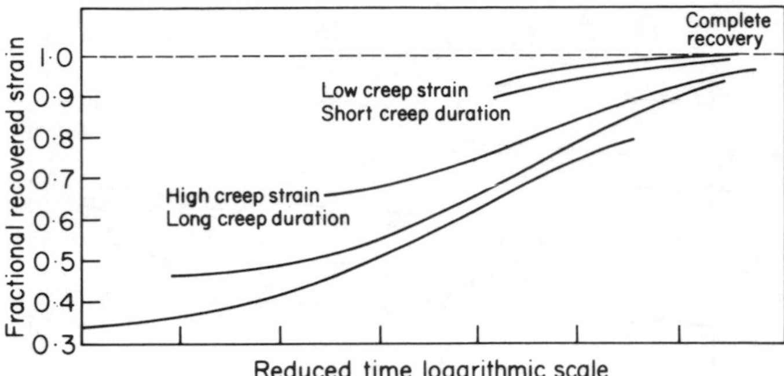

FIG. 6.15 Typical recovery behaviour. (Reproduced from BS 4618,* with permission of the British Standards Institution.)

Relevant standards and test methods

The standards relevant to creep deal mainly with general principles of testing and give recommendations rather than specifications for test apparatus and methods. They are ASTM D-674 (a 'Recommended Practices' specification), DIN 53 444 and BS 4618 : 1970 Part 1 Section 1.1.

The method of DIN 53 449 is partly relevant, although it is designed primarily for the determination of environmental stress cracking.

Methods and apparatus for creep testing have been reviewed by Ives, Mead and Riley.[10]

Typical creep data for reinforced thermoplastics

As has been mentioned (Chapter 1, Section 1.4, Chapter 4, Table 4.14), reinforcement, especially fibrous reinforcement, improves the resistance to creep of thermoplastic polymers. Some typical data are presented in Figs. 6.16–6.23.

Creep under stress can arise in various service situations, e.g. in pipes carrying a fluid (liquid or gas) under pressure, in stacks of plastics containers, crates etc. Applications of creep test data to practical design are discussed in materials manufacturers' publications and in the literature.[6,33–38]

Glass-fibre-reinforced thermoplastics have been ranked in the following order according to their ability to sustain constant loads at room temperature in creep experiments:

* 'Recommendations for the Presentation of Plastics Design Data', Section 1:1 1970 Creep. Copies of the complete standard may be obtained from the British Standards Institution, 2 Park Street, London W1A 2BS.

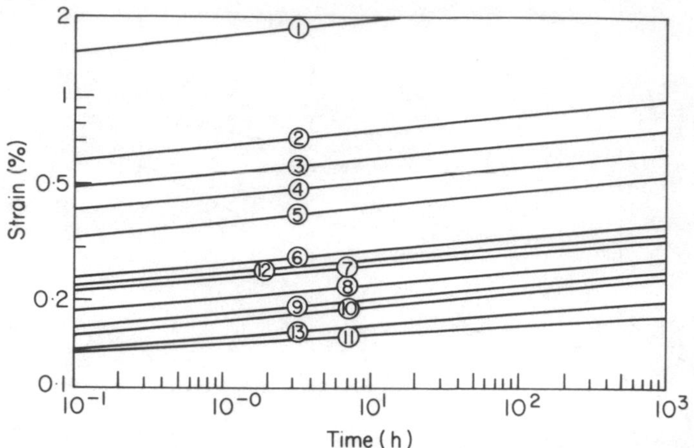

FIG. 6.16 Creep in tension of some engineering plastics: 23°C 50% RH. Stress = 150 kgf cm^{-2}. 1: N.6.6; 2: PPO mod.; 3: PC; 4: PET; 5: N.6+30% gf; 6: N.6.6+25% gf; 7: PET +18% gf; 8: N.6.6 +35% gf; 9: PPO mod. +30% gf; 10: PC +30% gf; 11: PET +30% gf (Arnite A340); 12: PBT +18% gf (Arnite T); 13: PBT +35% gf (Arnite T); gf = glass fibre. (Reproduced, with permission, from Technical Bulletins *Arnite T* and *Arnite 340* of AKZO Plastics BV.)

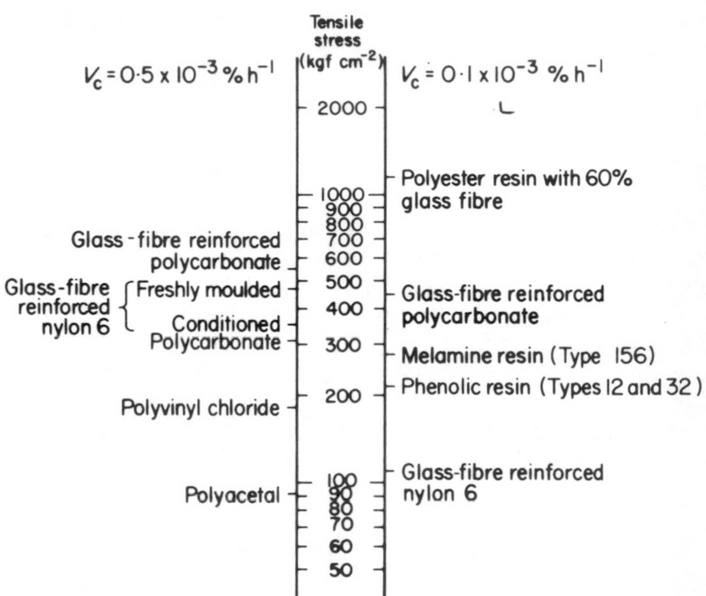

FIG. 6.17 Stress levels for the same creep rates in some plastics materials. The stress values shown are those which resulted in the average creep rates (V_c) of either $0{\cdot}5 \times 10^{-3}$% per hour (left-hand side) or $0{\cdot}1 \times 10^{-3}$% per hour (right-hand side). (Data of Streib and Oberbach:[31] from *Kunststoffe*, reproduced by permission of Carl Hanser Verlag.)

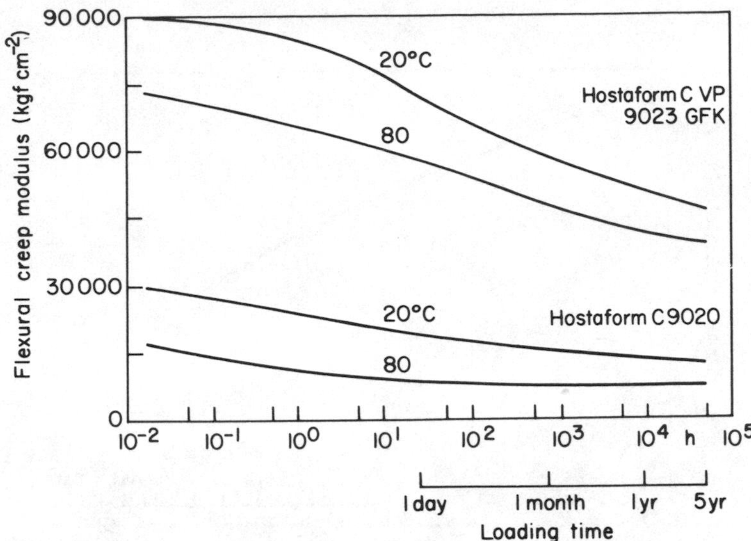

FIG. 6.18 Flexural creep modulus *vs* time: acetal copolymer with and without glass-fibre reinforcement. Hostaform C9020—general purpose injection and extrusion grade of acetal copolymer. Hostaform C VP 9023 GFK—the same polymer reinforced with 30% glass fibre. N.B. The current version of this grade is Hostaform 9023 GV (*cf.* Chapter 2, Table 2.2) which contains 25% glass fibre, but has better creep properties due to improved fibre-to-matrix bonding and large fibre length. Bending stress (edge fibre stress) 100 kgf cm^{-2}. (Reproduced, with permission, from the technical literature of Farbwerke Hoechst AG.)

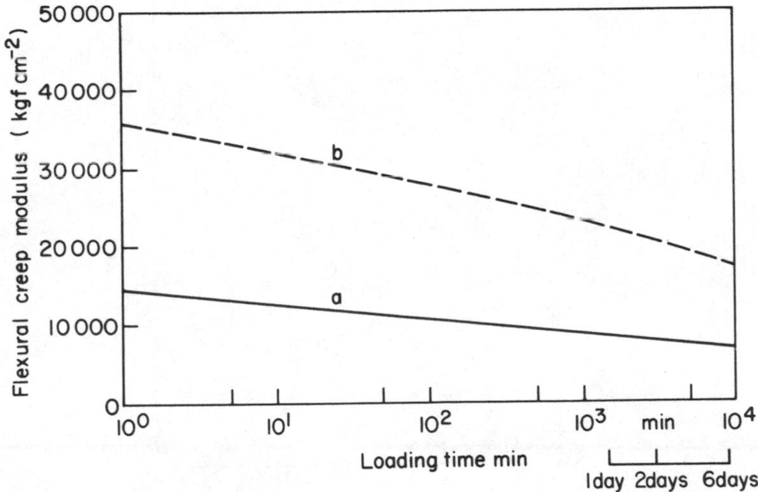

FIG. 6.19 Flexural creep modulus *vs* time: polypropylene unfilled, and reinforced with asbestos fibre; 20°C, bending stress 30 kgf cm^{-2} (a) Hostalen PPT 1070. (b) Hostalen PPT VP 7090 AV (40% asbestos fibre). (Reproduced, with permission, from the technical literature of Farbwerke Hoechst AG.)

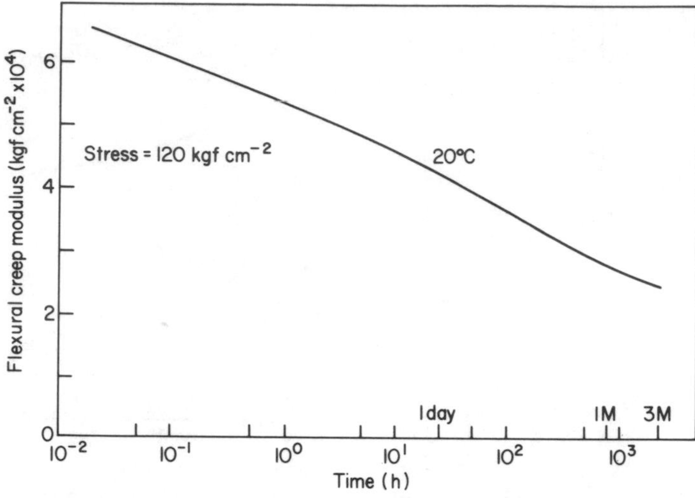

FIG. 6.20 Flexural creep modulus *vs* time: 30% glass-fibre-filled polypropylene (Hostalen PPN VP 7790 GV1). (Reproduced, with permission, from the technical literature of Farbwerke Hoechst AG.)

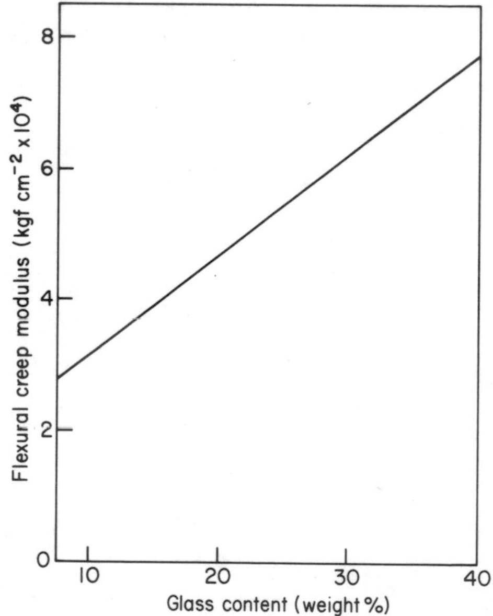

FIG. 6.21 Flexural creep modulus (1-min value) of glass-filled polypropylene (Hostalen PPN VP 7790 GV1) *vs* glass content. (Reproduced, with permission, from the technical literature of Farbwerke Hoechst AG.)

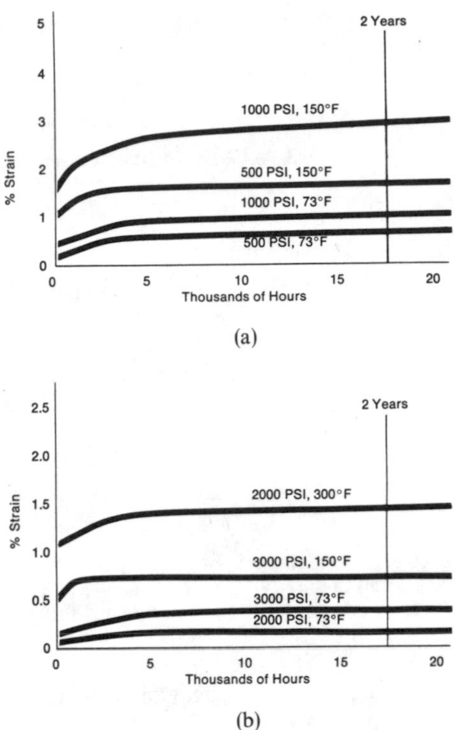

FIG. 6.22 Flexural creep of a thermoplastic polyester (PBT). (a) Unreinforced (Valox 310). (b) Reinforced with about 30% glass fibre (Valox 420). (Reproduced, with permission, from Technical Bulletin VAL-15A (Valox) of the General Electric Company.)

40% by weight glass fibre, 5000 psi: polystyrene = rigid PVC = polysulphone = polycarbonate > nylon 6.6 > polyacetal > nylon 6.10 > polypropylene. The first four exhibited a negligible creep rate after 1000 h.[39]

20% by weight glass fibre, 2000 psi: SAN > polystyrene > ABS > nylon 6 > polypropylene > polyethylene. The apparent moduli of the first four composites were found independent of stress level up to 5000 psi.[40]

Stress intermittently applied can also give rise to creep. In such situations there is usually partial recovery during the stress-free period of each cycle. Generally in thermoplastics the total creep strain after a number of cycles is lower than that produced by stress of the same magnitude applied continuously over the whole period.[6] Predictions of cyclic creep behaviour may be made from test data.[37,41] Such tests can

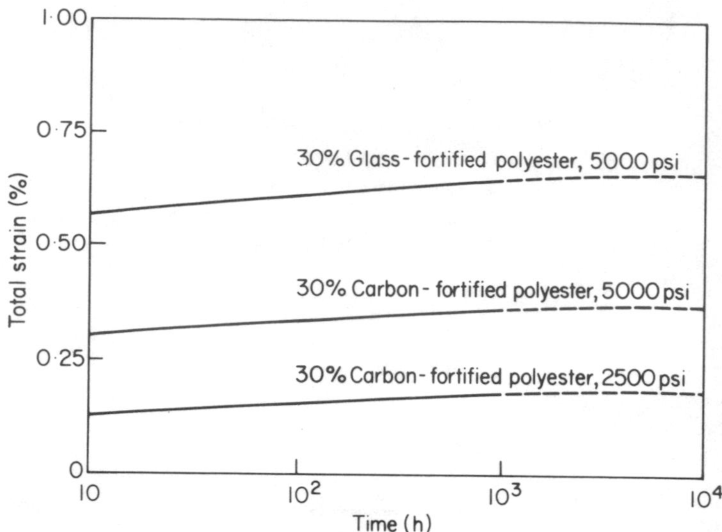

FIG. 6.23 Flexural creep of a thermoplastic polyester reinforced with carbon fibres. (Data of Theberge *et al.*;[32] Proceedings of the 29th Annual Conference RP/C Institute, SPI, 1974, reproduced with permission of the authors.)

also demonstrate not only the main effect of reinforcement (reduction of creep) but also that of improved surface adhesion (coupling) on the apparent modulus. With glass-fibre-reinforced, surface-coupled polypropylene and SAN, Cessna[41] found that intermittent creep strains increased very slowly after the first few cycles with a high degree of recovery after each cycle. The response of the same materials, but non-coupled, in identical conditions was radically different: the increase of the intermittent creep strain was monotonic and its rate high, with comparatively little recovery.

6.2.1.2.2 STRESS RELAXATION

This is the change in stress with time at constant strain: according to a definition in a relevant standard (ASTM D-674), 'the time-dependent change in the stress which results from the application of a constant strain to a specimen at constant temperature. The stress relaxation at a given elapsed time is equal to the instantaneous stress resulting when the strain is applied, minus the stress at the given time'.

In a determination the temperature and other relevant conditions are kept constant. The specimen is deformed rapidly and then held at constant strain. The stress is determined as a function of time and the results usually presented as a plot of stress *vs.* log time.

Stress relaxation modulus can be calculated from the results (ratio of stress to strain). This modulus is more useful than the creep compliance for some theoretical purposes. It should be noted that it is not the exact reciprocal of creep compliance (even for very small strains when both functions are linear) and hence is not equal to the 'creep modulus', although—except in severe transition regions—the difference is negligible for practical purposes.[5] Both moduli may be used (interchangeably) in design calculations.[6]

The practical difficulties of determination of stress relaxation increase with increasing stiffness of the test material, because the differences between the deformation of the specimen and that of the test apparatus may become small.[10]

The already mentioned ASTM D-674, and DIN 53 441(E) are two standards relevant to the determination of stress relaxation.

Tensile stress relaxation data for some glass-fibre-reinforced thermoplastics are given in Table 6.1.

6.2.1.2.3 FATIGUE

Fatigue in thermoplastics has been broadly defined[42] as '...the progressive weakening of a test piece or component with increasing time under load, such that loads which are satisfactorily accommodated at

TABLE 6.1
Tensile Stress Relaxation of Some Glass-fibre Reinforced Thermoplastics at Room Temperature
(Data of Theberge;[39] reprint by permission of *Modern Plastics International*, McGraw-Hill Inc.)

Base resin	Glass content by wt (%)	Applied stress (psi)	Decrease in applied stress (%)		
			1 h	5 h	15 h
Polycarbonate	40	15 000	10·7	13·0	14·7
Polysulphone	40	15 000	9·3	12·0	13·3
Nylon 6.10	60	15 000	18·7	25·0	28·3
Nylon 6.10	40	15 000	22·7	28·7	32·0
Nylon 6.6	60	15 000	14·7	19·3	22·0
Nylon 6.6	40	15 000	16·5	22·0	25·0
Polystyrene	40	10 000	11·0	15·0	15·0
Polycarbonate	40	10 000	10·0	12·0	12·0
Polysulphone	40	10 000	6·5	9·0	10·0
Polyacetal	40	10 000	20·0	25·0	28·0
Nylon 6.10	40	10 000	15·0	20·0	25·0
Nylon 6.6	40	10 000	12·5	16·0	20·0
PVC	35	10 000	14·3	18·5	20·5

short times produce failure at long times'. The subject has been reviewed by Andrews.[43,44]

In relation to thermoplastics, fatigue failure occurring under steady, continuous load (sometimes referred to as 'static fatigue'), is correctly termed *creep rupture* or *stress rupture.*

Fatigue resulting from the application of a periodically varying (cyclic) load is referred to as *dynamic fatigue* or simply 'fatigue'. In plastics it can be aggravated by temperature rises under cyclic loading, due to the high mechanical hysteresis and low thermal conductivity of these materials. In certain conditions failure can actually be caused by heat-softening (thermal fatigue).

Stress rupture

The modes of deformation employed in stress rupture tests on thermoplastics are usually tensile or flexure. Creep test apparatus of various types may be used, and the tests are simpler than creep tests, because accurate measurement of small deformations is not necessary.[5,10] Where possible tests may be carried out directly on moulded parts, to ensure that the effects of orientation (including reinforcing fibre orientation), and surface effects, are properly reflected in the results. The latter are normally presented as plots of stress *vs.* failure time (both, or at least the latter, usually on the log scale).

Progressively increasing loads may also be employed to determine the endurance limit (the Prot method[45]).

Perhaps the most important practical context in which the stress rupture properties of plastics have been considered and tested is the bursting of pipes under pressure. Suitable tests are given in standard specifications (BS 3505 and BS 3506, ASTM D-2239 respectively for PVC and polyethylene pipes); test methods and results for many plastics materials have been reviewed by Goldfein.[46]

The presence of glass-fibre reinforcement in a thermoplastic can alter the mode of failure in stress rupture at room temperature from ductile to brittle[47] and increase substantially the breaking stress both at room and elevated temperatures.[31]

Dynamic Fatigue

In a standard specification for testing flexural fatigue in plastics (ASTM D-671) fatigue is defined as 'the process of progressive localised permanent structural change occurring in a material subjected to conditions which produce fluctuating stresses and strains at some point or points and which may culminate in cracks or complete fracture after a sufficient number of fluctuations'.

Flexure is a popular deformation mode in fatigue tests on reinforced thermoplastics, although there are no fundamental reasons against other modes, e.g. tension or compression. The tests determine the *fatigue endurance,* the data being presented in the form of plots of stress (or stress amplitude) *vs.* the number of cycles endured or cycles to failure on a log scale. The plots are sometimes referred to as 'Wöhler curves'.

A useful flexural test technique has been described by Cessna, Levens and Thomson.[48] These authors have shown that the previously mentioned heating effect ('autogenous' or 'dissipative' heating) is the main cause of high-frequency flexural fatigue failure in several thermoplastic materials, unreinforced and reinforced. The modes of failure were drastic loss of stiffness or semi-ductile rupture; in isothermal conditions, with the specimen temperature kept down to the test temperature value, brittle failure occurred. The presence of glass-fibre reinforcement (30% by weight) in polypropylene, ABS and SAN resulted in substantial improvements in fatigue endurance, especially with effective interfacial adhesion (coupling).

Typical results for some thermoplastics with and without reinforcement are shown in Figs. 6.24–6.27.

6.2.2 Other Properties

As has already been pointed out in the present chapter and elsewhere, reinforcing fillers modify not only the mechanical properties, but also other

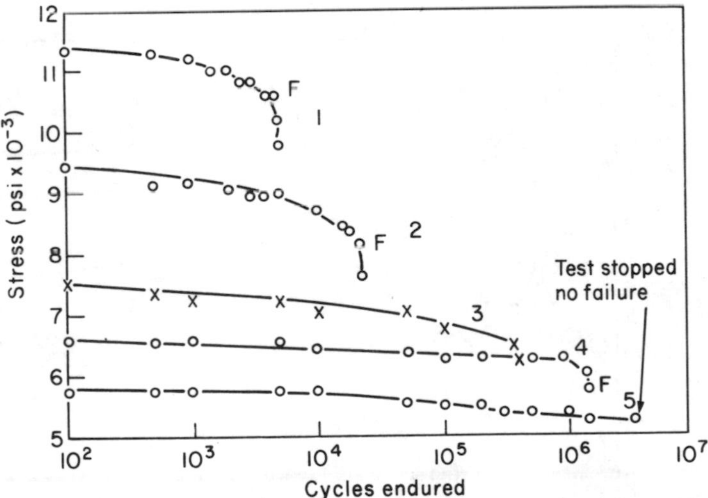

FIG. 6.24 Apparent stress *vs* cycles endured for glass-reinforced and unreinforced thermoplastics. (1) Polypropylene with 30% glass fibre (coupled); (2) polypropylene with 30% glass fibre; (3) SAN with 30% glass fibre; (4) ABS; (5) polypropylene. (Data of Cessna Jr. *et al.*; Proceedings of the 24th Annual Conference RP/C Division, SPI, 1969, Paper 1-C, reproduced by permission of the authors.)

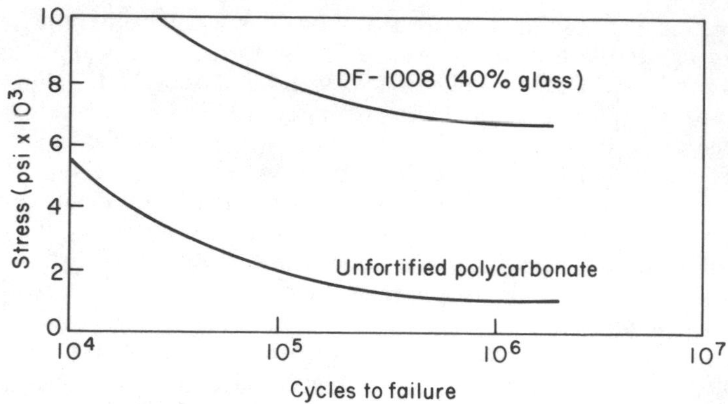

FIG. 6.25 Fatigue endurance of polycarbonate (glass-filled: Thermocomp DF 1008—LNP).

properties of thermoplastic polymers. Whilst these latter modifications are incidental in the sense that they are not the primary object of incorporation of reinforcement, their extent may be substantial: the density, for example, may be increased by a factor of 1·5 or more. Electrical and thermal properties may be considerably affected, as may frictional

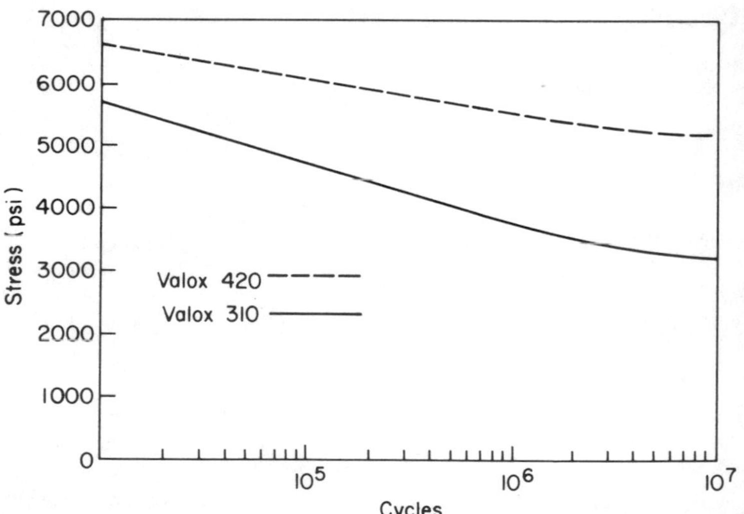

FIG. 6.26 Fatigue endurance of thermoplastic polyester (PBT). Unreinforced: Valox 310; reinforced: Valox 420 (30% glass fibre). (Reproduced, with permission, from Technical Bulletin VAL-15A (Valox) of the General Electric Company.)

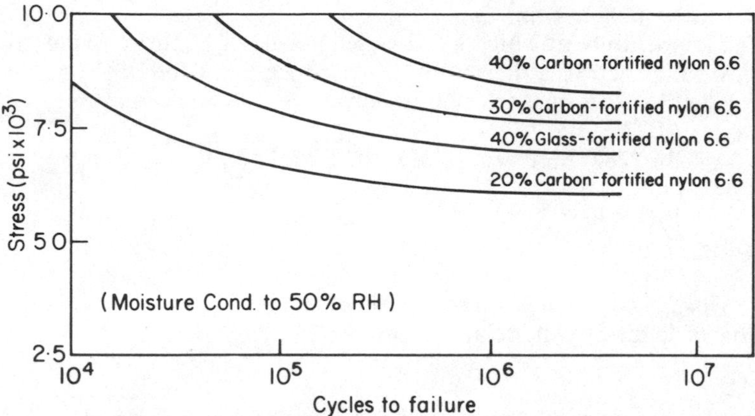

Fig. 6.27 Fatigue endurance of nylon 6.6 reinforced with carbon fibre and glass fibre. (Data of Theberge *et al.*; Proceedings of the 29th Annual Conference RP/C Institute, SPI, 1974, reproduced with permission of the authors.)

properties. Optical properties of clear polymers are invariably influenced, because of the opacifying effect of reinforcing fillers.

In this section those non-mechanical properties of reinforced thermoplastics which are of the greatest practical interest are defined by reference to relevant standard tests (Section 6.2.2.1) and some typical values are quoted (Section 6.2.2.2).

6.2.2.1 Standard Definitions and Test Methods

The form and order of presentation of the definitions and other information in this section are the same as in the section on mechanical property tests (see above).

6.2.2.1.1 THERMAL PROPERTIES

(i) *Heat distortion temperature**

Temperature at which a prescribed arbitrary deformation of a defined test specimen occurs under a defined set of testing conditions.

Tests: ASTM D-648; ISO R75
BS 2782 (102 G and 102 H)
DIN 53 461

Units: °F or °C for maximum fibre stress in the specimen of: (a) 18·5 kgf cm^{-2} (264 psi) and/or (b) 4·6 kgf cm^{-2} (66 psi).

* Also known, from the title of ASTM D-648, as 'deflection temperature under load'.

(ii) *Vicat softening point*

The temperature at which a flat-ended needle of 1 mm^2 circular cross-section will penetrate a thermoplastic specimen to a depth of 1 mm under a specified load, using a selected uniform rate of temperature rise.

Tests: ASTM D-1525
BS 2782 (Methods 102 D, 102 F and 102 J)
DIN 53 460
ISO R306

Units: °F or °C

(iii) *Coefficient of linear thermal expansion*

This is defined by the relationship:

$$\alpha = \Delta L / L_0 \Delta T$$

where α = coefficient of linear thermal expansion per degree Fahrenheit or Celsius; ΔL = change in length of test specimen due to heating or cooling; L_0 = length of test specimen at the reference temperature (room temperature); ΔT = temperature difference (°C or °F) over which ΔL is measured.

Tests: ASTM D-696
BS 4618 (Section 3.1)
VDE* 0304 (Parts 1 and 4)

Units: in in^{-1} $°F^{-1}$; in in^{-1} $°C^{-1}$; cm cm^{-1} $°C^{-1}$

(iv) *Thermal conductivity*

Quantity of heat per unit time passing normally through a unit area of a sheet of material of unit thickness for a unit temperature difference across the faces,[10] in steady conditions.

Tests: ASTM C-518
ASTM C-177
BS 874 (Clause 10c)
DIN 52 612

Units: Btu in ft^{-2} h^{-1} $°F^{-1}$; cal cm cm^{-2} s^{-1} $°C^{-1}$; J m m^{-2} s^{-1} $°C^{-1}$; or W m^{-1} $°C^{-1}$

(v) *Thermal diffusivity*

Thermal diffusivity is equal to (and may be calculated from the determined values of)

$$\frac{\text{thermal conductivity}}{\text{specific heat} \times \text{density}}$$

This physical parameter may be regarded as the thermal analogue of the

* German Electrical Engineers Association.

diffusion coefficient, and has the same dimensions and units ($cm^2\ s^{-1}$; $m^2\ s^{-1}$). It is relevant to all problems of transient heat flow.

A simple, accurate and rapid *direct* method of determining thermal diffusivity has been devised by Braden.[49]

6.2.2.1.2 ELECTRICAL PROPERTIES

(i) *Volume resistivity*

The electrical resistance between opposite faces of a unit cube: it may also be defined in terms of potential gradient and current density (*cf.*, e.g. ASTM D-257).

Tests: ASTM D-257
BS 2782 (Method 202A)
IEC Publ.* P3 and 167

Units: Ω cm; Ω m

(ii) *Surface resistivity*

Surface resistivity may be defined as the resistance between surface-mounted electrodes of unit width and unit spacing (with volume currents assumed minimal).[10] Definitions in terms of potential gradient and current per unit width of surface are also used (*cf.*, e.g. ASTM D-257).

Tests: ASTM D-257
BS 2782 (Method 203A)
IEC Publ. 93

Units: Ω

(iii) *Permittivity*

Strictly *relative permittivity* (previously dielectric constant). Defined as the ratio of the capacitance (C_x) of a given configuration of electrodes with a material as the dielectric, to the capacitance (C_v) of the same electrode configuration with air (more strictly vacuum) as the dielectric, i.e. $k = C_x/C_v$.

Tests: ASTM D-150
BS 2782 (Method 207A)
DIN 53 483

Units: None (a ratio)

(iv) *Dissipation factor*† (*loss tangent*‡)

The tangent of the loss angle or the cotangent of the phase angle (for full definitions see Ref. 10 or ASTM D-150, or IEC Publication 250).

* International Electrotechnical Commission, Geneva.
† Term favoured in the USA and Europe.
‡ Term favoured in the UK.

Tests: ASTM D-150
BS 2782 (Method 207A)
DIN 53 483
Units: None (a ratio)

(v) *Dielectric strength*

Usually defined in terms of the field strength (ratio of voltage to thickness) required to produce breakdown of the material in specified conditions (*cf.* also ASTM D-149).

Tests: ASTM D-149
BS 2782 (Method 201A)
DIN 53 481
Units: V mil^{-1}; kV mm^{-1}; V m^{-1}

(vi) *Arc resistance*

Time to break-down of (formation of a conducting path in) the surface of a material under the action—in specified conditions—of an arc of high voltage and low current produced between two surface-mounted electrodes.

Tests: ASTM D-495
DIN 53 484
BS 3497 (not normally used with plastics)
Units: s

6.2.2.1.3 ABRASION (WEAR) AND FRICTION

Abrasion

This is usually measured in terms of weight or volume of material abraded away from the surface of a material in arbitrary, rigidly specified conditions. Light scattering by the surface of the specimen is also sometimes used as an index of this property. The Taber method and apparatus (ASTM D-1044) are sometimes employed with reinforced thermoplastics.

Tests: ASTM D-1044 (intended for transparent plastics)
ASTM D-1242
Units: Appropriate to method, e.g. g or mm^3 lost per number of cycles

The results of standard tests for abrasive wear resistance are not directly applicable to the behaviour in service of the materials of bearings, gears and the like. Thermoplastics, e.g. nylon, acetal polymers, fluoropolymers, are used in this type of application, and their usefulness increases with improved mechanical and physical properties conferred by reinforcement, providing that the wear and frictional properties are suitable.

Tests are normally run to determine wear factors, in apparatus relevant to applications of this kind. This subject is further discussed in Section 6.2.2.2 below.

Friction

Friction between two surfaces in contact is measured in terms of the static and dynamic (kinetic) coefficients of friction, μ_s and μ_D, defined respectively by the expressions:

$$\mu_s = F_s/L \quad \text{and} \quad \mu_D = F_D/L$$

where F_s = the minimum force required to initiate sliding; F_D = the minimum force required to sustain steady sliding; L = the force (load) acting normally on the surfaces.

Tests: ASTM D-1894
BS 2782 (Method 311A)

Units: None (a ratio)

Several non-standard methods of friction measurement have been described.[10] Friction and abrasive wear are interrelated: the load sustained by the surfaces in contact is a factor in both these phenomena. The principal factors of volumetric wear in thermoplastics are represented by the expression:

$$W_v \propto \frac{Ld \tan \theta}{H}$$

where W_v = volumetric wear; L = load; $\tan \theta$ = roughness factor (asperity angle); H = indentation hardness of the plastics surface; d = sliding distance.

6.2.2.2 Effects of Reinforcement

6.2.2.2.1 THERMAL PROPERTIES

The incorporation of reinforcing fillers in thermoplastics reduces thermal expansion and mould shrinkage. Quantitative data on the effects in particular cases are given in the Tables of Section 3.3, Chapter 3.

The softening point (e.g. Vicat) is essentially a property of the matrix and is only moderately affected by even fairly substantial reinforcement loading, or by the nature of the filler. This is illustrated by the following figures for nylon 12 and polypropylene (commercial materials from Chemische Werke Hüls A. G.).

The melting point of crystalline base polymers (e.g. polyacetal, nylon 6.6, polypropylene) is unaffected by reinforcement.

The deflection temperature under load, which has been classified with the mechanical properties in this chapter, can be raised substantially by the incorporation of reinforcing filler (*cf.* Fig. 6.4 above, tables in Section 3.3 Chapter 3, and Table 4.14 Chapter 4).

Polymer	Filler (% by weight)	Vicat softening point (ISO R 306)	
		5 kg load	1 kg load
Polypropylene (Vestolen P 5200)	—	95	155
Polypropylene (Vestolen P 5232G)	Glass fibre approx. 30%	100	160
Polypropylene (Vestolen P 5232T)	Talc approx. 30%	100	160
Nylon 12 (Vestamid L1901)	—	138–142	172–175
Nylon 12 (Vestamid L1930)	Glass fibre approx. 30%	156–170	175–176
Nylon 12 (Vestamid X2363)	Glass spheres approx. 30%	149	173

In a paper discussing the practical significance of heat distortion temperature (deflection temperature under load) in reinforced thermoplastics Krautz[50] points out that whilst this property is strongly influenced by the reinforcement content it is also affected by the morphology of the base polymer. In this respect the significant factors appear to be the glass temperature and the crystalline melting point of, respectively, glassy and crystalline polymers.

Thermal conductivity can be affected, the magnitude of the effect depending on the nature of the base polymer, that of the reinforcement, and amount incorporated. For example, the thermal conductivity of nylon 6 can be increased by a factor of about 2·5 on incorporation of 30% glass fibre by weight (from 1·3 to 3·3 Btu in ft^{-2} h^{-1} $°F^{-1}$*) whilst that of nylon 12 is reduced from 0·25 to 0·21 kcal m^{-1} h^{-1} $°C^{-1}$†. The thermal conductivity of carbon-fibre filled thermoplastics can be greater by factors of 10 to 15 than that of corresponding glass-fibre filled compounds.[51]

6.2.2.2.2 ELECTRICAL PROPERTIES

To the extent to which generalisation is possible, the general effects of the main types of reinforcement on the electrical properties of reinforced thermoplastics may be summarised as follows.

In some materials (e.g. SAN, acetal copolymers) the electrical properties of the base polymer are little affected by incorporation of glass or asbestos fibres. In others (e.g. nylon 6 and 6.6) the dielectric strength is increased, whilst arc resistance may or may not be modified (*cf.* Table 6.2). The effects

* Unreinforced nylon 6 and Thermocomp PF 1006 respectively.
† Vestamid L1901 and L1930 respectively.

TABLE 6.2

Electrical Properties of Some Thermoplastics With and Without Reinforcement

Base polymer	Filler: Nature	Filler: Wt %	(ASTM D-257) Volume resistivity (Ω cm)	(ASTM D-150) Permittivity (60 Hz–10^6 Hz)	(ASTM D-150) Dissipation factor (60 Hz–10^6 Hz)	(ASTM D-149) Dielectric strength (V mil^{-1})	(ASTM D-495) Arc resistance (s)
Polystyrene	—	0	$>10^{16}$	2·4–2·7	0·0001–0·0004	500–700	60–140
Polystyrene	'Short' glass fibre	30	$>10^{16}$	2·8(5)–2·8(5)	0·0007–0·0008	500–600	90–110
Polypropylene	—	0	$>10^{16}$	2·2–2·6	0·0004–0·0018	500–660	136–185
Polypropylene	'Long' glass fibre	40	$\geq 10^{16}$	2·5–2·5	0·002–0·004	250–350	60–80
Polypropylene	Asbestos fibre	40	10^{15}–10^{16}	2·7–2·6	0·0070–0·0020	440–450	115–125
Polypropylene	Talc	40	$\geq 10^{16}$	2·6–2·6	0·0050–0·0060	—	—
Modified PPO	—	0	$>10^{16}$	2·6(5)–2·6(5)	0·0004–0·0009	400–550	70–80
Modified PPO	'Short' glass fibre	30	$\geq 10^{17}$	2·9–2·9	0·0009–0·0015	550–600	100–120
Polycarbonate	—	0	$>10^{16}$	3·2–2·9	0·0006–0·0100	390–410	110–120
Polycarbonate	'Short' glass fibre	30	10^{15}–10^{16}	3·5–3·4(5)	0·0010–0·0075	450–480	110–120
Nylon 6.6[a]	—	0	10^{15}–10^{16}	4·1–3·6	0·026–0·024	360–370	130–140
Nylon 6.6[a]	Glass fibre	30	10^{15}–10^{16}	4·0–3·8	0·007–0·017	400–480	125–135
Nylon 6.6[a]	Asbestos fibre	30	10^{12}–10^{13}	4·1–3·5	0·032–0·017	440–460	125–130

[a] Dry as moulded.

on volume and surface resistivities depend on the nature and amount of the reinforcement, and vary somewhat in different base polymers. In the majority, the resistivities are increased—in varying degrees—by the addition of glass or asbestos fibres, and reduced (drastically at higher loading levels) by carbon fibres. Theberge *et al.*[32] report surface resistivity values of the order of 1 Ω for a number of thermoplastics containing 30% carbon fibres.

The effects on loss tangent (dissipation factor) and permittivity (dielectric constant) vary with the polymer and the type and nature of reinforcement. In general, in glass and asbestos reinforced thermoplastics, the effects are consistent with an increase in capacitance on incorporation of the fillers. The permittivities and dissipation factors of several thermoplastics vary with the frequency; among those least sensitive in this way, within the range 60–10^6 Hz, are glass-reinforced polystyrene and modified PPO.[52]

The effects of reinforcement on the electrical properties of some thermoplastics are shown in Table 6.2.

6.2.2.2.3 WEAR AND FRICTION

As has been mentioned, the main interest in these properties of reinforced thermoplastics is in connection with their use in bearings, gears and other applications where low-friction base polymers (nylon, acetal, fluoropolymers) have been used alone.

The performance of plastics bearing materials is usually assessed in terms of the relationship between the radial wear (reduction in wall thickness) of a bearing and the 'PV value', the product of specific bearing load (load per unit projected area) and peripheral velocity of the shaft journal. The relationship may be expressed by the equation[53]

$$r = KPVT \tag{6.1}$$

in which r is the radial wear (for a unidirectionally loaded cylindrical fixed bearing this is approximately equal to volumetric wear divided by the projected area).

K is the proportionality constant in a simple version of the expression of the type given in Section 6.2.2.1.3, *viz.*:

$r \propto Ld$

$r = KLd$

$\quad = KLVT$

where V and T are respectively the relevant velocity and running time, i.e. $d = VT$, P is the load per unit projected area, i.e. $P = L/A$.

A rearrangement of eqn 6.1 to:

$$r/T = KPV \tag{6.2}$$

provides the basis for plotting *PV* against unit wear in a given time (often 1000 h).[53] A typical plot is shown in Fig. 6.28.

The '*PV* limit' or 'limiting *PV*' is the highest acceptable value of *PV* consistent with satisfactory sustained operation of a bearing (at constant bearing temperature and coefficient of friction).

Wear and friction data for some base polymers alone in comparison with the same polymers lubricated by the addition of PTFE, and also both glass-fibre reinforced and lubricated, are given in Table 6.3, and other, more general data in Fig. 6.29. Lubricated, reinforced thermoplastics are further discussed in Chapter 9, Section 9.3.

In general, the incorporation of glass-fibre reinforcement tends to increase the friction coefficients of base polymers, and hence reduce the limiting velocity in bearings. The wear of the bearing as a whole also tends to be increased: the mechanism whereby this occurs is comparatively complex,[55] but the effect is basically due to the abrasive nature of the glass. It is for this reason that glass-reinforced thermoplastics are often modified to improve their frictional properties (*cf.* Table 6.3 and Section 9.3 Chapter 9). Carbon fibres, especially the high modulus type, can improve the wear performance of a thermoplastic bearing or gear material, whilst at the same time improving the mechanical properties.[51,55] However, the effect is also influenced by the nature of the polymer matrix and the interfacial adhesion.[56]

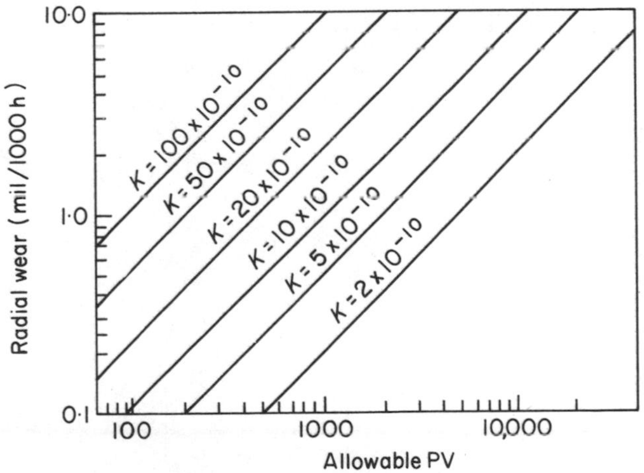

FIG. 6.28 Radial wear *vs* PV as a function of wear factor. Sample: cylindrical sleeve bearing, fixed, loaded unidirectionally. Test conditions: cold rolled carbon steel shaft, 12–16 micro-inch finish and 18–22 Rc, room temperature, dry. (Data of Theberge:[53] reprint by permission of *Modern Plastics International*, McGraw-Hill Inc.)

TABLE 6.3

Limiting PV, Wear Factor, and Coefficient of Friction of Internally Lubricated Thermoplastic Composites
(Data of Theberge:[53] reprint by permission of *Modern Plastics International*, McGraw-Hill Inc.)

Base resin	Glass (wt %)	TFE (wt %)	Limiting PV (journal half bearing) (10 ft min^{-1})	(100 ft min^{-1})	(1000 ft min^{-1})	Wear factor[c] (equilibrium) (10^{-10} in^3 min ft^{-1} lb^{-1} h^{-1})	Coefficient of friction: Static (40 psi)	Dynamic (40 psi, 50 ft min^{-1})
Styrene-acrylonitrile, BFL-4036[a]	30	15	17 500	10 000	10 000	65	0·13	0·18
Polycarbonate, D-1000	0	0	750	500	NR	2 500	0·31	0·38
Polycarbonate, DFL-4036	30	15	27 500	30 000	13 000	30	0·18	0·20
Polysulphone, GFL-4036	30	15	20 000	35 000	15 000	70	0·09	0·11
Acetal, K-1000	0	0	4 000	3 500	<2 500[b]	65	0·14	0·21
Acetal, Fulton-404	0	20	10 000	12 500	5 500	17	0·07	0·15
Acetal KFL-4036	30	15	12 500	12 000	8 000	200	0·20	0·28
Polypropylene, MFL-4036	30	15	14 000	12 000	7 500	36	0·09	0·09
Nylon 6, PFL-4036	30	15	17 500	20 000	13 000	17	0·20	0·25
Nylon 6.10, QFL-4036	30	15	20 000	15 000	12 000	15	0·23	0·31
Nylon 6.6, R-1000	0	0	3 000	2 500	>2 500	200	0·24	0·26
Nylon 6.6, RL-4040	0	20	14 000	27 500	8 000	12	0·10	0·18
Nylon 6.6, RFL-4036	30	15	17 500	20 000	13 000	16	0·19	0·26
Polyurethane, TFL-4036	30	15	7 500	10 000	5 500	35	0·20	0·25
Polyester, WFL-4036	30	15	17 000	19 000	12 000	20	0·16	0·21

[a] LNP Thermocomp series code no.
[b] Low load limit of test apparatus at 1000 ft min^{-1}.
[c] Wear factor K values are for 50 ft min^{-1}, 40 psi, 2000 PV. A material with a wear factor of $K = 1$ will wear 1 in^3 supporting a load of 1 lb at a velocity of 1 ft min^{-1} in 1 h.

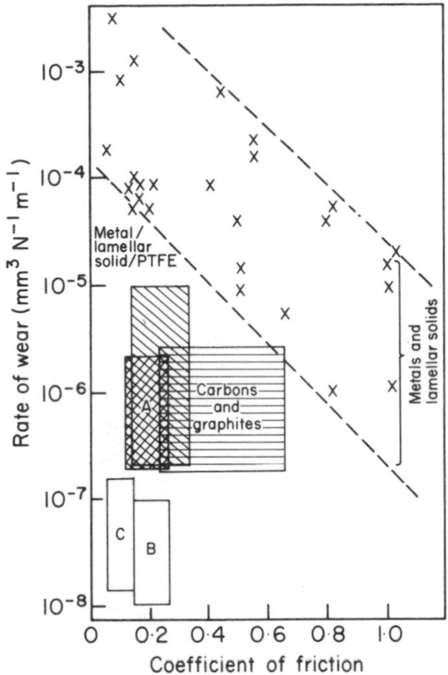

FIG. 6.29 Comparative tribological properties of composites. (A) Thermoplastics/carbon fibre or glass fibre; (B) thermosets/carbon fibre; (C) woven glass fibre/PTFE fibre/resin and porous bronze/PTFE/lead. (Data of Giltrow;[54] Crown copyright, reproduced from *Composites* with the permission of the Controller H.M. Stationery Office.)

6.3 EFFECT OF TEMPERATURE ON THE PROPERTIES OF REINFORCED THERMOPLASTICS

The properties of reinforced thermoplastics are affected by temperature within the range of practical interest to the designer and user because those of the base polymers are temperature-sensitive in that range. The effects on mechanical and physical properties are particularly important in practice.

The effects of temperature may be either:

(a) essentially physical (largely reversible); or

(b) chemical, irreversible, resulting in permanent modification (e.g. degradation, cross-linking).

The first type of effect may be particularly pronounced and abrupt if the temperature change spans a transition point (e.g. glass temperature, crystalline melting point). The second may occur, in an essentially innocuous environment, solely or mainly as a result of heating at elevated temperature, or it may take place at normal acceptable temperatures in the presence of chemically active environments or radiation. Some environmental effects are considered further in Section 6.4.

6.3.1 Temperature Effects on Mechanical Properties

These are of great significance because of their importance in engineering applications.

6.3.1.1 Short-term Properties

There are two aspects of the effect of temperature on short-term properties of thermoplastic materials which are important in use and service. Firstly, a rise or drop in temperature sufficiently short not to produce any permanent chemical changes can nevertheless alter the physical state of the material sufficiently to affect the properties. Single-point measurement of properties at, say, room temperature and an elevated temperature will give different results. The second aspect is that of a permanent change in the properties due to chemical effects when the material is raised to, and maintained at, an elevated temperature that is sufficiently high. Here the effect is essentially one of time at a particular temperature and the property change, which usually increases with time, is irreversible.

At temperatures significantly above room temperature the mechanical properties of thermoplastics, especially strength and stiffness, are normally reduced. Presence of reinforcing fillers decreases the magnitude of this effect. The stiffness and strength generally tend to increase as the temperature is lowered, although the stress/strain response and the mechanism of failure may also change in consequence of a temperature drop.

The immediate effects of temperature on key mechanical properties of some reinforced thermoplastics in comparison with unreinforced grades are illustrated in Figs. 6.30–6.34.

The thermal endurance of a thermoplastic, i.e. the effect of elevated temperature over long periods of time (*cf.* Figs. 6.35 and 6.36) is important to the designer and user.

What might be called the 'immediate' properties at the particular elevated temperature can be evaluated in terms of single-point tests, but it is obviously important to be able to predict the maximum service temperature which is 'safe' for a particular thermoplastic material's term of expected service. A useful practical embodiment of this concept is the 'temperature index'[58]—a temperature limit for continuous service for a particular material—predicted on the basis of heat ageing procedures devised by the Underwriters' Laboratories Inc.*

* Further information on this organisation is given in Chapter 9 Section 9.2.1.

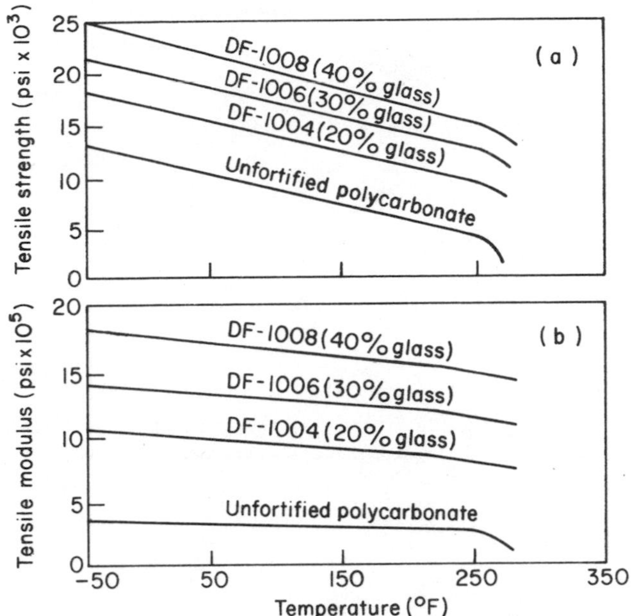

FIG. 6.30 Effect of temperature on: (a) tensile strength; (b) tensile modulus, of polycarbonate. (Thermocomp DF Series; courtesy of the LNP Corp.)

The tests, which are arbitrary (but standardised), amount to an evaluation of the effects of controlled thermal degradation on a few key properties, commonly tensile strength, impact strength and dielectric strength. The property changes on ageing are followed over a long period, at a number of elevated temperatures. Log time to failure at a number of temperatures is plotted against the reciprocal of absolute temperature; extrapolation of the plot gives a value for the thermal endurance at room temperature, and this is used in conjunction with a plot for a control material, to calculate the temperature index.[58] The method is based on the premise that thermal degradation of thermoplastics is an activation process governed by a reaction-rate relationship of the Arrhenius type. The generally good correlation of predictions from the test results with effects of actual long-term service provides supporting evidence for the validity of the method.

Materials for which the correlation has been fully established in the light of sufficient experience receive a generic index. Other indices apply to specific grades or products actually tested. The indices are published by the Underwriters' Laboratories Inc. Most are included in the yearly edition of the *Modern Plastics Encyclopedia* (McGraw-Hill Inc.).

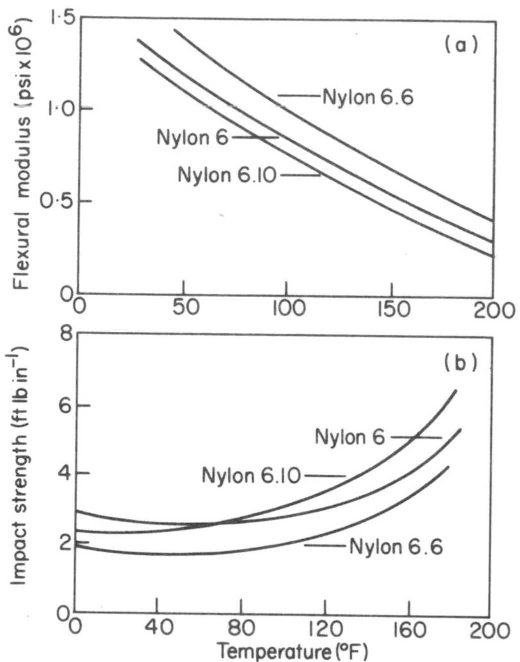

FIG. 6.31 Some temperature effects on the properties of 30% glass-fortified nylon resins. (a) Flexural modulus; (b) notched Izod impact strength. (Courtesy of the LNP Corp.)

Examples of temperature index values are given opposite (figures current in 1971): the index may be different for different properties, because the effects of temperature may be different.

6.3.1.2 LONG-TERM PROPERTIES

In general the resistance of thermoplastics to creep, stress rupture and fatigue decreases, and stress relaxation increases with rising temperature. These effects are reduced, in some cases substantially, by the incorporation of reinforcing fillers: see, e.g. Figs. 6.18 and 6.22 for illustration of creep at room and elevated temperature of two polymers—acetal copolymer and PBT—filled and unfilled.

Test data obtained at elevated temperatures over comparatively short time intervals are sometimes used to predict long-term behaviour at room temperature. That is, the effects of temperature are treated as equivalent to those of time. Such time-temperature superpositions are limited in applicability, but they may be used in some cases to provide design information.[5,6,59] Stress-time superposition has also been applied at various

Material	Temperature index (°C)
Nylon 6.6, 6 and 6.10 (with and without glass-fibre reinforcement)	65 (generic)
Polycarbonate (with and without glass-fibre reinforcement)	65 (generic)
Zytel 7010–33 (glass-fibre-reinforced nylon 6.6 moulding compound)	
minimum material thickness 0·125 in, resistance to impact not essential	120
minimum material thickness 0·031 in, only electrical properties essential	105
Lexan 3412 (glass-fibre-reinforced polycarbonate moulding compound)	
minimum material thickness 0·062 in	120
minimum material thickness 0·062 in, resistance to impact not essential	130
Lexan 101 (medium-viscosity polycarbonate moulding compound)	
minimum material thickness 0·062 in	115
minimum material thickness 0·062 in, resistance to impact not essential	125

temperatures, *inter alia* to glass-reinforced polypropylene:[60] general rules and limitations of this procedure have been formulated by Turner.[5] Prediction techniques for creep behaviour of polyethylene are discussed by Moore and Turner.[61]

Some typical examples of changes in long-term mechanical properties of thermoplastics with increasing temperature, and of improvements brought about by reinforcing fillers, are given in Figs. 6.37–6.40.

6.3.2 Temperature Effects on Other Properties

The electrical properties of thermoplastics can vary with temperature, generally like the mechanical properties, i.e. both as a result of differences in physical state of the material at different temperatures and also because of permanent degradation (*cf.* Section 6.3.1). Incorporation of reinforcing fillers often reduces the magnitude of the change in short-term effects and also modifies the effect of degradation because it can make the material more resistant to that. In some polymers the incorporation of reinforcing fillers makes comparatively little difference to the effect of temperature on electrical properties (*cf.* Figures 6.41 and 6.42).

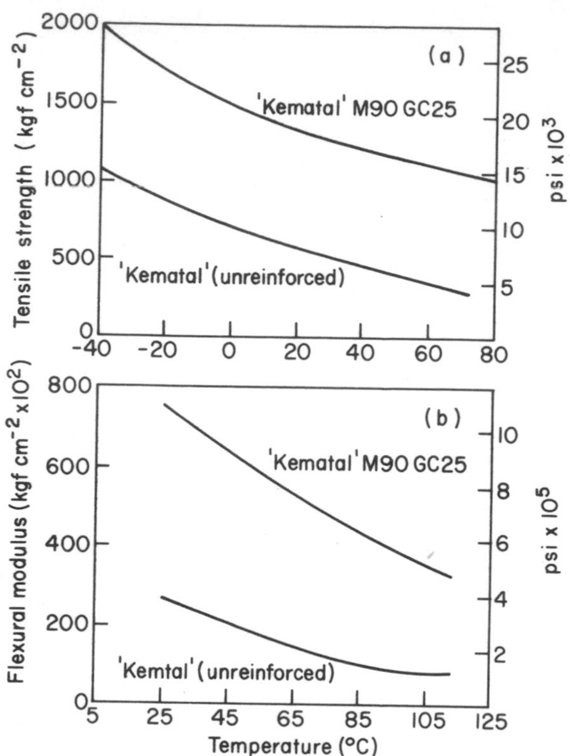

FIG. 6.32 Acetal copolymer (Kematal^R): (a) tensile strength (yield); (b) flexural modulus, as functions of temperature. Kematal M90 GC25 contains 25% glass fibre by weight. (Courtesy of Amcel Ltd.)

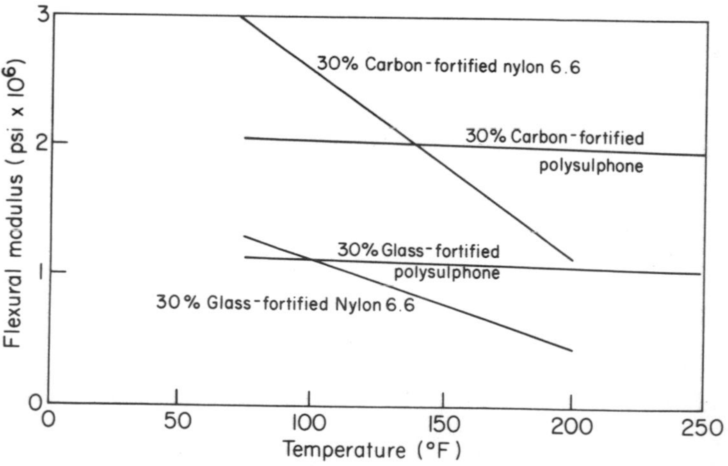

FIG. 6.33 Flexural modulus of some reinforced thermoplastics as a function of temperature. (Data of Theberge *et al.*;[32] Proceedings of the 29th Annual Conference RP/C Division, SPI, 1974, reproduced by permission of the authors.)

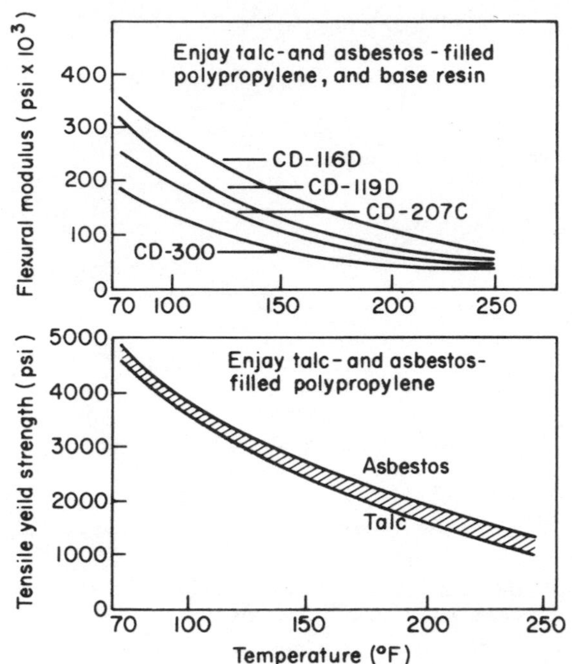

FIG. 6.34 Filled polypropylene; effect of temperature on: (a) stiffness; (b) tensile strength; CD-116D: 40 wt % asbestos: CD-119D: 40 wt % talc; CD-207C: 22 wt % talc; CD-300: base resin. (Reproduced, with permission, from the Technical Bulletin *Reinforced Polypropylene.* PLA-69-1622 of the Exxon Chemical Co., USA.)

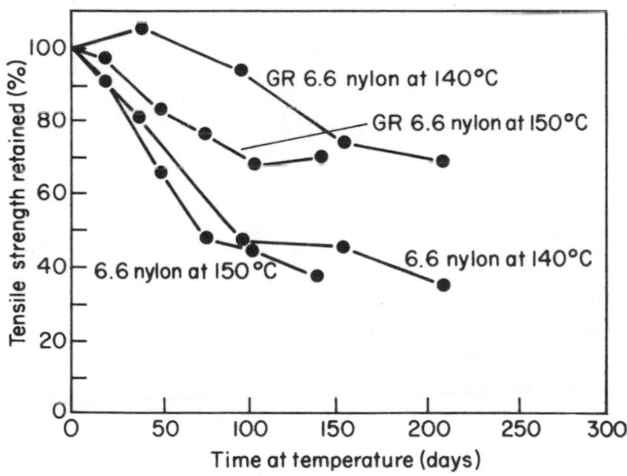

FIG. 6.35 Effect of time at elevated temperature on heat-stabilised nylon 6.6 with and without glass-fibre reinforcement. (Data of Theberge *et al.*;[57] reproduced, with permission, from *Machine Design,* The Penton Publishing Co.)

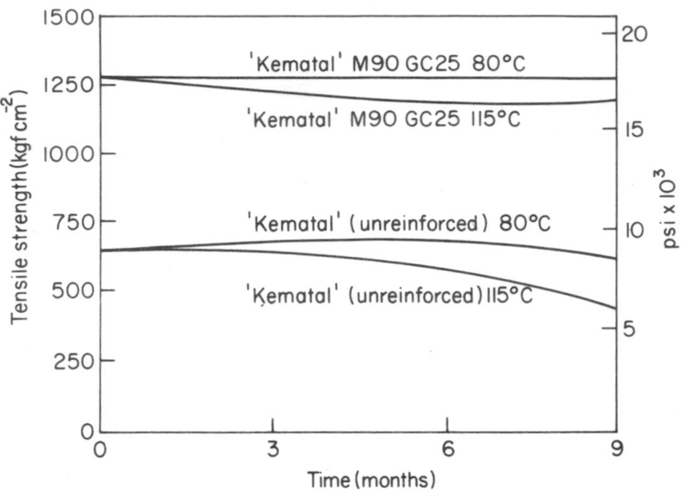

FIG. 6.36 Acetal copolymer (Kematal[R]): effect of heat ageing on tensile strength. Materials as in Fig. 6.32 (Courtesy of Amcel Ltd.)

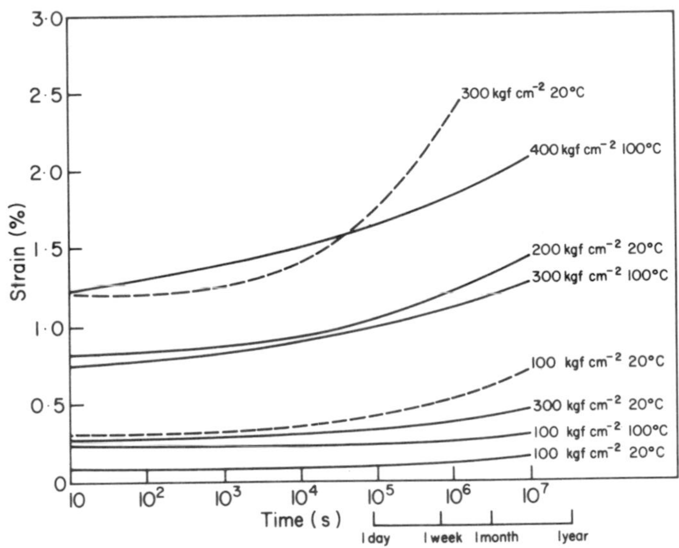

FIG. 6.37 Effect of temperature on creep of a thermoplastic polyester (PBT: Deroton) unfilled and with glass-fibre reinforcement. (---- unreinforced; —— glass reinforced (30 wt %)). (Based on data abstracted, with permission, from TS Notes PMC TD 201 and PMC TD 202, ICI Plastics Division.)

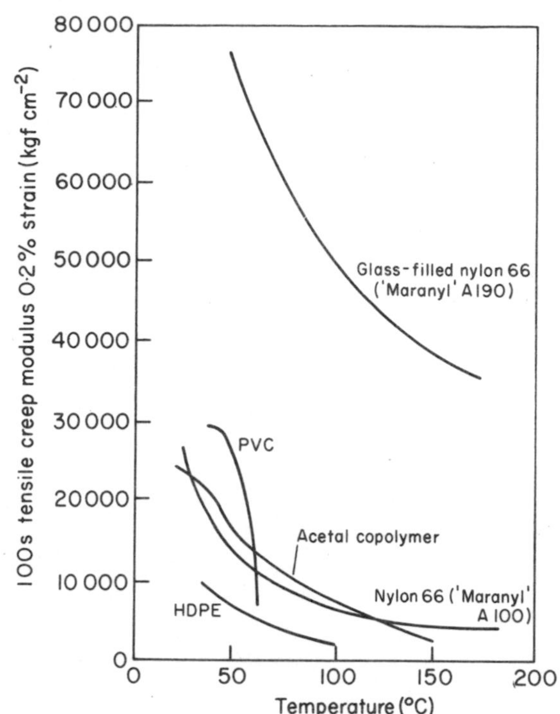

FIG. 6.38 Tensile creep modulus (100 s 0·2% strain) *vs* temperature. Dry nylon 66, glass-filled nylon 66 (Maranyl A100, A190) and other polymers. (Reproduced, with permission, from TS Note N104 (second edition), ICI Plastics Division.)

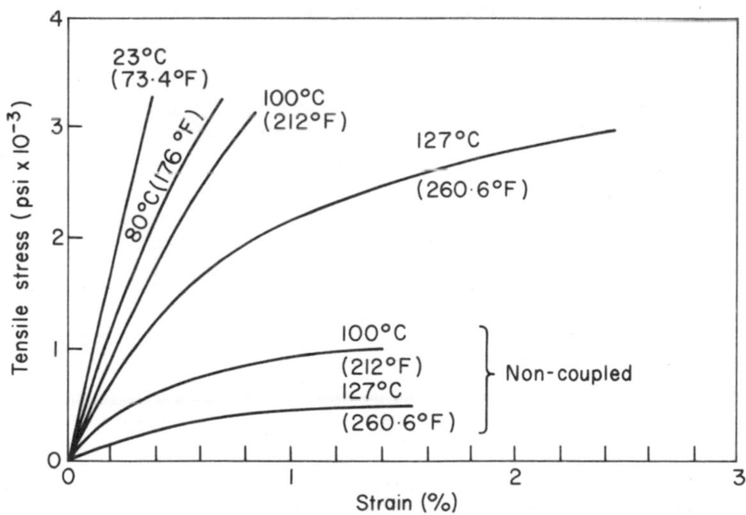

FIG. 6.39 Stress–strain curves for 30 wt % 'coupled', glass-fibre-reinforced polypropylene at varous temperatures, obtained during 300 s loading. Curves for non-coupled materials are included for comparison. (Data of Cessna: reprinted, with permission, from the *SPE Journal*, **28,** February 1972.)

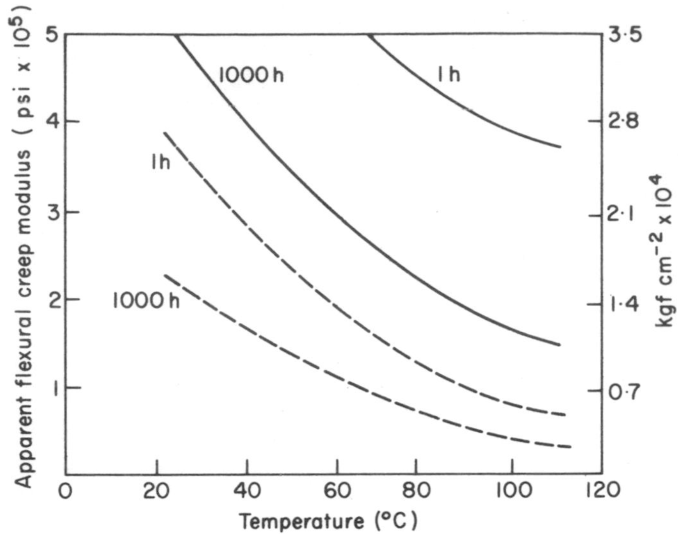

FIG. 6.40 Acetal copolymer: effect of temperature on apparent modulus at 1 h and 1000 h. (---- unreinforced polymer; —— polymer with 20 wt % glass fibre.)

Frictional properties normally deteriorate with rising temperature, as does the performance of reinforced thermoplastics in bearings and sliding parts.

As is known, the thermal conductivity of thermoplastics can vary with temperature (*cf.* Fig. 6.43). This variation can be influenced by incorporation of reinforcement.

In the British Standard 'Recommendations for the Presentation of Plastics Design Data'[7] Section 2.1 includes recommendations on a suggested layout for a data sheet representing variation of permittivity with temperature and frequency.

6.4 ENVIRONMENTAL EFFECTS

The effects of environment on polymeric materials are often grouped under the three main headings of 'chemical effects', 'weathering' and 'stress crazing and cracking'. This kind of division, often convenient within the context of a particular treatment, is nonetheless artificial because each of the three groups includes phenomena occurring as a result of chemical and physical factors or interactions of the two. Thus, for example, deterioration of a polymer through weathering may be a complex effect of oxidation, chain scission and cross-linking through radiation effects and absorption of water which may act as a swelling agent. Cracking occurring as a result of weathering may outwardly closely resemble environmental stress cracking produced by the combined effect of the presence of internal stresses in a moulding and the absorption of a swelling agent. The effects of time and temperature already discussed also become superimposed on those just mentioned.

For the purpose of the discussion in the present section, the main environmental factors will be considered in conjunction with, and in the order of the nature and magnitude of, their effects upon thermoplastics and in particular reinforced thermoplastics. The classification, whilst completely arbitrary, is convenient and no more artificial than the more

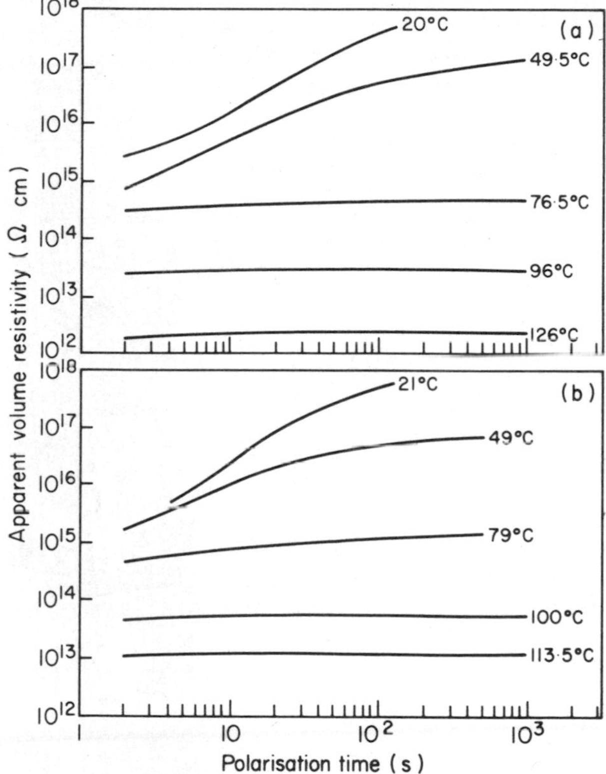

FIG. 6.41 Temperature dependence of apparent volume resistivity of dry thermoplastic polyester. (a) Deroton TAP 10 (ICI polybutylene terephthalate, unfilled). (b) Deroton TGA 50 (ICI polybutylene terephthalate 30 wt % glass fibre). (Reproduced, with permission, from TS Notes PMC TD 201 and PMC TD 202, ICI Plastics Division.)

conventional sub-grouping of environmental effects. Thus, in the order of increasing severity, the environmental effects may be listed as:

(a) *Modification of properties without actual failure: in the broadest sense of the term this may be called 'plasticisation'*

This can be brought about by absorption of swelling agents (reagents which do not cause dissolution or degradation). The effects can be very similar to the immediate effects of a rise in temperature (*cf.* Section 6.3)

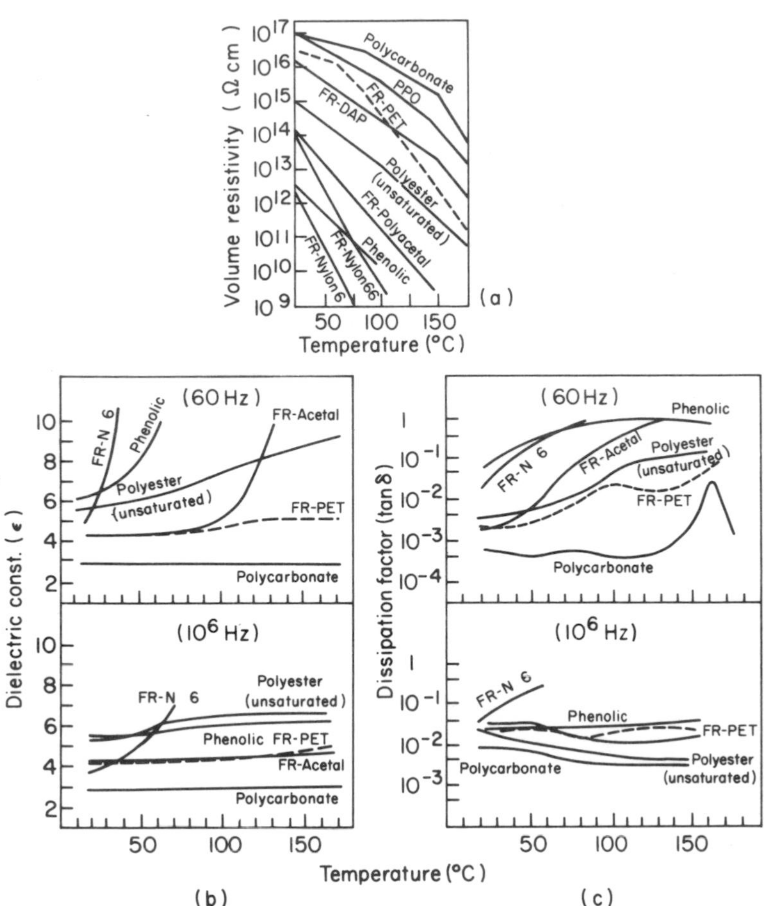

FIG. 6.42 Effects of temperature on some electrical properties of thermoplastics in comparison with thermoset phenolic and polyester materials. FR: glass-fibre reinforced. (a) Electrical volume resistivity *vs* temperature; (b) dielectric constant (ε) *vs* temperature; (c) dissipation factor (tan δ) *vs* temperature. (Reproduced, with permission, from the technical bulletin FR-PET of Teijin Ltd.)

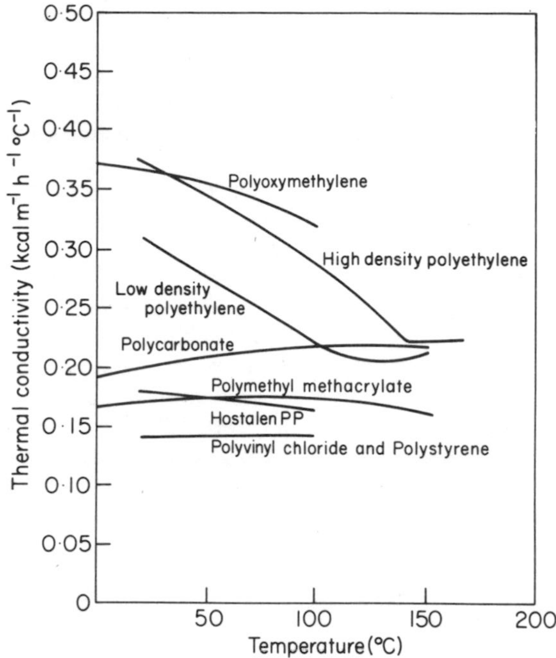

FIG. 6.43 Thermal conductivity of some thermoplastics as a function of temperature. Hostalen PP: Hoechst polypropylene. (Reproduced, with permission, from the technical literature of Farbwerke Hoechst AG.)

because, like plasticisation by penetrants, this can increase the freedom of molecular movement: the effect may be substantial and fairly sharp if the change spans the glass transition point.

(b) *Drastic modification of properties involving what might be called 'partial failure' of varying severity*

The two main limiting cases of this are:

(i) damage arising from essentially the physical effect of 'plasticisation' (in the widest sense) by a penetrant in the presence of stress, i.e. stress crazing or cracking; and

(ii) damage, which may be similar to (i) in outward appearance, caused e.g. by oxidation or chain rupture, often as a result of exposure to atmospheric oxygen and/or radiation possibly aggravated by the presence of moisture or atmospheric pollutants. These effects are thus essentially chemical in nature.

The type of failure just mentioned under (i) and (ii) above has been referred to as 'partial' because it does not involve complete breakdown or extensive loss of substance and the material retains its overall shape.

(c) *Disintegration or dissolution of the material (or, in the case of a reinforced thermoplastic, of the polymer matrix)*
The mechanisms of this effect are essentially chemical.

6.4.1 'Physical' Modification of Properties ('Plasticisation Effects')

Experience and tests indicate that, in general, reinforced thermoplastics, and in particular glass-reinforced materials, tend to be more resistant to solvents than the corresponding base polymers alone.[57] However, reagents which can penetrate a thermoplastic polymer will be sorbed also when that polymer is the matrix in the reinforced thermoplastic with the resultant modification of properties and swelling if enough penetrant has been sorbed. In some cases, particularly where interfacial adhesion between the polymer matrix and reinforcement is not good, and where the surface of the reinforcement is readily wetted by the penetrant, sorption may actually be enhanced as e.g. through wicking along reinforcement fibres.

Exposure of reinforced thermoplastics to penetrants (vapour, or liquid

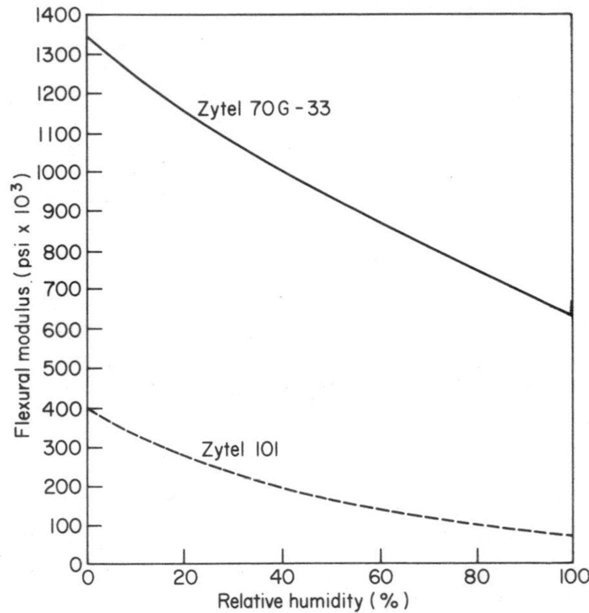

FIG. 6.44 Nylon 6.6: flexural modulus *vs* humidity at 23°C (specimens at equilibrium). Zytel 70G-33: 33% glass fibre by weight; Zytel 101: base polymer alone. (Reprinted, with permission, from Technical Bulletin TRZ 6909, Du Pont de Nemours International SA.)

by immersion) can lead to changes in properties which, though they may differ in extent, will be generally similar in nature to those produced in the base resin alone. Information on the latter effects is available from the general literature and manufacturers' data on the particular thermoplastic polymers. The usual practice is to describe the effects either qualitatively, e.g. as 'slight', 'fair', etc., or in terms of quantitative differences in a measured property, or a combination of both. This does not distinguish between any plasticising effects and those of partial chemical degradation.

Because of its abundance and distribution, water—either as liquid or vapour in the atmosphere—is the most common potential penetrant encountered by reinforced thermoplastics. It is sorbed in significant proportions by some base polymers of reinforced thermoplastics, notably nylons (*cf.* Chapter 3), resulting in effective plasticisation with an attendant effect on physical properties. Other properties can also be strongly affected, e.g. the dielectric properties.

Ageing in water or in the presence of appreciable amounts of moisture can lead to permanent changes in properties. This is mentioned in Section 6.4.2. The effect of moisture sorption on a short-term mechanical property (the stiffness) of nylon 6.6, with and without reinforcement, is illustrated in Fig. 6.44. The effect of moisture content on electric breakdown strength of reinforced nylon 6.6 is shown in Fig. 6.45, and the effect on a long-term mechanical property (creep) is illustrated by Fig. 6.46.

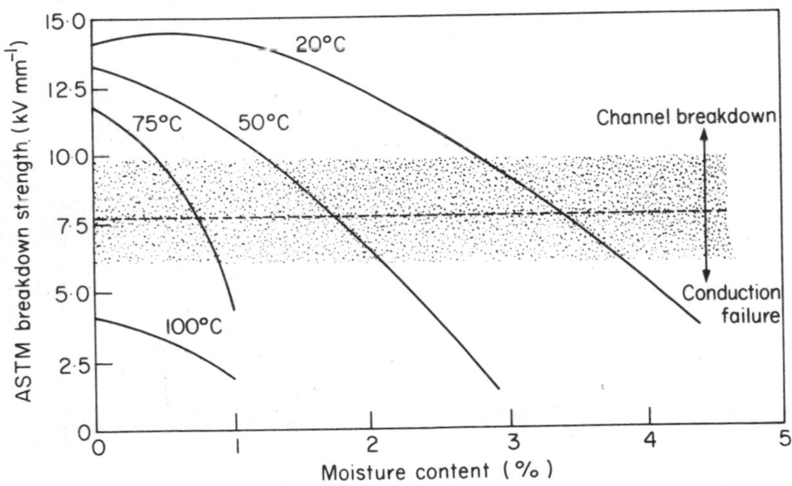

FIG. 6.45 Variation of electric breakdown strength of Maranyl A190 (glass-filled) nylon with water content (at 20°C) for different temperatures. (Reproduced, with permission, from TS Note N110—3rd edition, ICI Plastics Division.)

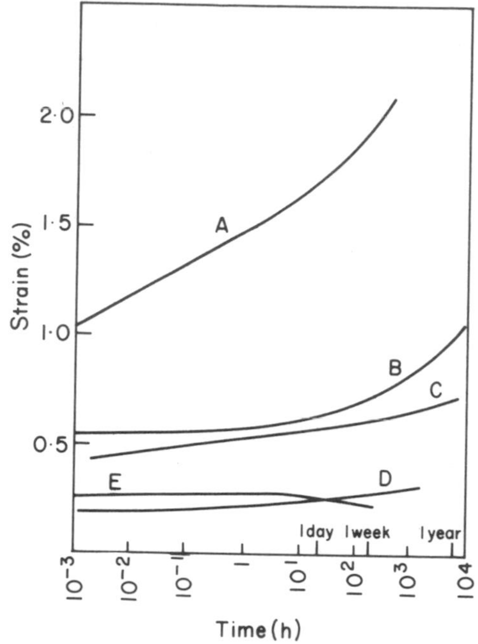

FIG. 6.46 Creep in tension: 20°C, comparison of wet and dry specimens. Glass-filled nylon 6.6 (Maranyl A190). (A) Maranyl A190, 100% rh 500 kgf cm^{-2}; (B) Maranyl A190, 0% rh 500 kgf cm^{-2}; (C) Maranyl A190, 100% rh 200 kgf cm^{-2}; (D) Maranyl A190 0% rh 200 kgf cm^{-2}; (E) die-cast aluminium alloy LM2. (Reproduced, with permission, from TS Note N104, ICI Plastics Division.)

6.4.2 Permanent Modification and/or Partial Failure

As has been mentioned, two types of effect can be distinguished here: partial failure and/or permanent modification as a result of chemical changes in the polymer matrix caused by such environmental factors as oxygen, radiation (including sunlight) or other 'weathering' agencies, and environmental stress cracking.

6.4.2.1 Chemical Modification

As already noted, this tends to be slower and less drastic in reinforced thermoplastics than in the base polymers alone, but otherwise the effects of environment in this respect are similar in kind. Reduction in strength, stiffness and other mechanical properties can result, with accompanying changes in other properties, e.g. electrical. A frequent manifestation is weathering, which may include the complex effects of some or all of the main common factors, *viz.* oxidation, hydrolysis and photodegradation, giving rise to chain scission, formation of reactive groups and sites, and cross-linking. These effects may be promoted or accelerated by microbiological attack, presence of penetrants (e.g. water) and heat, and, conversely, environmental factors can accelerate or increase the effects of temperature on thermoplastics materials. A useful account of weathering tests has been published by Caryl.[62]

In some polymers sensitive to hydrolysis the presence of glass-fibre reinforcement can promote the operation of this mechanism. In glass-reinforced polyacetal and polysulphone this has been attributed to enhanced penetration by wicking along the fibres and greater reaction at the fibre/polymer interface.[57] Some effects of ageing in hot water are illustrated by the data of Table 6.4.

Data on the chemical resistance and related properties of reinforced thermoplastics, including compatibility with surface finishes (*inter alia* metal plating of ABS and polypropylene) have been published by Theberge, Arkles and Cloud.[40,57]

6.4.2.2 Environmental Stress Crazing and Cracking

The two phenomena may together be described generally as environmental stress failure. They are in fact forms of stress rupture brought on or accelerated by an environment active in this sense. Environmental stress failure does not involve appreciable chemical modification of the polymer. Crazing is a characteristic local, essentially plastic, deformation of a material to beyond the critical strain, resulting in the formation of microvoids but no large-scale discontinuity. Cracking is essentially local brittle failure. Either or both can occur appreciably below the limits of the polymer's normal short-term strength in air, in the presence of stress (externally imposed or internal moulding stresses) in a fluid environment (gas vapour or liquid) which acts as a cracking or crazing agent. Crazed material is load-bearing but its mechanical properties are generally inferior to those of the solid polymer.[43,63] Cracks are usually wedge-shaped and grow at the tip where the stress tends to be concentrated. The critical strain for the formation of crazes or cracks may be defined as that minimum strain at which under a given set of conditions (and in particular in the presence of a given crazing or cracking agent) stress crazing or cracking is known to start. In addition to the fluid environment, temperature and time (in effect the rate of strain) form the 'set of conditions' in this definition.

Two main general explanations have been advanced for the mechanism of environmental stress crazing and cracking. According to one the cracking or crazing agent reduces, by wetting, the energy for craze formation and thus promotes and facilitates growth of holes (craze cells and cracks) from minute voids in the polymer which serve as nuclei. The second

TABLE 6.4

Effect of Water Ageing at 100°C on Tensile Strength of Glass-Fortified Thermoplastics

(Data of Theberge *et al.*:[57] reproduced, with permission, from *Machine Design*, The Penton Publishing Co.)

Base resin	Glass content		Tensile strength (psi) Time (h)					
	Wt (%)	Vol (%)	0	100	1 000	5 000	10 000	50 000
SAN	30	15·4	16 700	9 500	6 000	3 000	—	—
Polycarbonate	30	16·8	19 800	10 000	5 600	4 000	—	—
	0	0·0	8 800	9 700	5 600	4 000	—	—
Polysulphone	30	17·3	17 100	12 000	12 200	12 600	12 800	12 900
	0	0·0	10 500	10 900	11 900	12 500	—	—
Polyacetal	30	19·2	14 100	10 000	8 000	2 500	—	—
	0	0·0	9 500	9 800	9 300	2 000	—	—
Polypropylene	30	13·2	7 600	7 000	6 600	6 200	6 000	5 850
Nylon 6.10	30	15·4	20 000	9 500	8 300	6 000	4 500	—
PVC	25	15·1	17 400	11 500	8 600	6 700	5 000	—
Modified PPO	30	15·2	21 500	18 000	14 000	14 000	13 500	5 000
	0	0·0	8 600	8 900	8 600	8 300	4 500	—

explanation, which currently appears to be the more strongly supported by available evidence, is the so-called plasticisation hypothesis. According to this the cracking or crazing agent penetrates and plasticises the polymer, reducing the critical strain at the craze or crack tip (where there is a concentration of stress) and thus facilitating the propagation of the craze or crack. Certain other specific mechanisms can operate in particular cases, but they are not generally applicable. For example, in bisphenol A polycarbonate morphological rearrangements including crystallisation can occur under the influence of certain penetrants, which can give rise to local stresses and crazing or cracking in the presence of the penetrant acting as crazing agent.[64] The cracking of nylon 6 under the influence of metal chlorides in aqueous or alcoholic solution was attributed (by Dunn and Sansom[65]) to local swelling caused by formation of a complex between the alcohol, the metal and the amide groups in the polymer. Whilst polycarbonate is capable of crystallisation, and nylon 6 is a crystalline polymer, in general environmental stress failure is less common in crystalline polymers than in amorphous polymers and is still less well understood, despite the fact that environmental stress cracking of another crystalline polymer—polyethylene—in contact with certain polar liquids, especially surface active agents, has been known since the mid-1940s and extensively studied.

The possible significance of hydrogen bonding as a supplementary factor in the stress crazing and cracking of amorphous thermoplastics by the plasticisation mechanism, has been discussed by Vincent and Raha.[66]

The literature on environmental stress failure phenomena in thermoplastics is extensive. The subject has been reviewed recently, both in its theoretical[43,67] and practical[68] aspects. Numerous methods have been proposed and used for the investigation of environmental stress failure of thermoplastic polymers, and/or the prediction of stress-cracking or crazing behaviour of mouldings in service.[68] Several of these methods are applicable to reinforced thermoplastics but there are no standard test methods specifically for these materials. Of the general standards available, three deal explicitly with the stress cracking of polyethylene in the form of standard specimens (ASTM D-1693 and D-2552) or blow-mouldings (ASTM D-2561). The procedure of the German Standard DIN 53 449, based on the work of Pohrt,[69] is suitable both for more fundamental stress cracking studies and for comparative evaluation of stress cracking in mouldings. The Pohrt method may be used with reinforced thermoplastics.

Practical studies to date of environmental stress cracking of reinforced thermoplastics, in particular glass-reinforced materials, have not revealed any pronounced improvement or deterioration in the stress cracking tendency of the polymer attributable directly and solely to the presence of the reinforcement. In general two factors are relevant here.

Local stress concentrations can arise in the neighbourhood of reinforcing fibres or particles or the ordinary moulding stresses can be enhanced through local differential cooling of the polymer matrix and reinforcement particle. Since the thermal expansion coefficients of the base polymers are much higher than those of the common reinforcements (*cf.* Chapter 4), the tendency is for the polymer to 'shrink-on' around the reinforcement when the moulding is cooled and this can set up stresses at the interface. Such stresses may then promote cracking in the presence of penetrants. An examination of the stress cracking behaviour of four polymers (polypropylene, SAN, acetal and polyamide) in various solvents, an oil, a detergent and bleach,[46] indicated no particular effects clearly attributable to either the presence or the percentage content (between 0 and 40% by weight) of the glass fibre reinforcement. Crazing of the polymer matrix has been put forward (by Lavengood and co-workers[70]) as the principal feature of a suggested deformational mechanism to account for unexpected differences in the toughness and ultimate elongation of polystyrene, SAN and PPO filled with glass beads.

6.4.3 Disintegration or Dissolution

These are strong chemical effects which may be accelerated by temperature. Because of the catastrophic effect on the base polymer, such slight reduction in the rate of deterioration as may be caused in some cases by the presence of reinforcement is not of real significance.

REFERENCES

1. Ogorkiewicz, R. M. and Weidmann, G. W. (1973). 'Properties of Reinforced Thermoplastics and Design', Paper presented at the Symposium on Advances in Reinforced Thermoplastics, April 13th, Kingston Polytechnic, Kingston-on-Thames, England.
2. Ogorkiewicz, R. M. (1971). *Composites,* **2,** 1, 29.
3. Ogorkiewicz, R. M. and Turner, S. (1971). *Plastics and Polymers,* **39,** 141, 209.
4. Lieng-Huang Lee and Ricke, J. K. (1969). 24th Conference, RP/C Division, SPI, Paper 1-E.
5. Turner, S. (1973). *Mechanical Testing of Plastics,* Plastics Institute and Iliffe Books, London.
6. Ogorkiewicz, R. M. (Ed.) (1974). *Thermoplastics: Properties and Design,* John Wiley and Sons, London.
7. British Standard 4618:1970; Recommendations for the Presentation of Plastics Design Data.
8. Taprogge, R. (1973) *Kunststoffe,* **63,** 7, 469.
9. Turner, S. *loc. cit.*, p. 106.

10. Ives, G. C., Mead, J. A. and Riley, M. M. (1971). *Handbook of Plastics Test Methods,* Iliffe Books, London.
11. Barrie, J. A. (1968). *Diffusion in Polymers,* (Ed. J. Crank and G. S. Park), Academic Press, London and New York, Chapter 8.
12. Braden, M. (1968). *J. Polym. Sci. Part A-1,* **6,** 1227.
13. Stafford, G. D. and Braden, M. (1968). *J. Dental Res.* **47,** 2, 341.
14. Fujita, H. (1961). *Fortschr. Hochpolym.-Forsch.* **3,** 1–47.
15. Crank, J. (1967). *The Mathematics of Diffusion,* Oxford University Press.
16. Meares, P. (1964). *Polymers: Structure and Bulk Properties,* Van Nostrand, New York.
17. Meares, P. (1966). *European Polymer Journal,* **2,** 95.
18. Barrer, R. M. (1968). *Diffusion in Polymers,* (Ed. J. Crank and G. S. Park), Academic Press, London and New York, Chapter 6.
19. Theberge, J. E. and Hall, N. T. (1969). 24th Conference, RP/C Division SPI, Paper 1-B.
20. Wilson, M. G. (1972). 27th Conference, RP/C Institute, SPI, Paper 11-E.
21. Maaghul, J. (1971). 26th Conference, RP/C Division, SPI, Paper 8-A.
22. Hollingsworth, B. L. (1969). *Composites,* **1,** 1, 28.
23. Lanham, B. J. (1972). 'Reinforced Thermoplastics', Paper presented to the North-Western Section of the Plastics Institute, Manchester, England, 24th February.
24. Alfrey, T. (1948). *Mechanical Behaviour of High Polymers,* Interscience, New York.
25. Gross, B. (1953). *Mathematical Structure of the Theories of Viscoelasticity,* Herman et Cic, Paris.
26. Ferry, J. D. (1970). *Viscoelastic Properties of Polymers,* 2nd edn, Wiley, New York.
27. Bateman, L. (Ed.) (1963). *The Chemistry and Physics of Rubber-like Substances,* Wiley, New York.
28. Kambour, R. P. and Robertson, R. E. (1972). In *Polymer Science,* (Ed. A. D. Jenkins), North-Holland, Amsterdam, Vol. 1, Chapter 11.
29. Pearson, J. R. A. *loc. cit.*, Chapter 6.
30. Joisten, S., *Die Maschine,* Nos. 10/68 and 1/69, A. G. T. Verlag, Ludwigsburg.
31. Streib, H. and Oberbach, K. (1965). *Kunststoffe,* **55,** 5, 309.
32. Theberge, J., Arkles, B. and Robinson, R. (1974). 29th Conference, RP/C Institute, SPI.
33. Streib, H. (1968). *Kunststoffe,* **58,** 2, 4 and 4, 7.
34. ICI Technical Service Notes K 109 (1968) and N 110 (1971), ICI Plastics Division, Welwyn Garden City, Herts, England.
35. Cessna, L. C., Jr. (1971). *Polym. Eng. Sci.,* **11,** 211.
36. Powell, P. C. (1973). *Plastics for Industrial Designers,* The Plastics Institute, London.
37. Powell, P. C. and Turner, S. (1971). *Plastics and Polymers,* **39,** 144, 261.
38. Lee, M. (1972). 27th Conference, RP/C Institute SPI, Paper 14-B.
39. Theberge, J. E. (1968). *Modern Plastics,* June, p. 155.
40. Theberge, J., Arkles, B. and Cloud, P. (1972). 27th Conference, RP/C Institute, SPI, Paper 14-C.
41. Cessna, L. C. (1972). *SPE Journal,* **28,** 2, 28.

42. Bucknall, C., Gotham, K. V. and Vincent, P. I. (1972). *Polymer Science,* (Ed. A. D. Jenkins). North Holland, Amsterdam, **1,** Chapter 10.
43. Andrews, E. H. (1973). *The Physics of Glassy Polymers,* (Ed. R. N. Haward), Applied Science Publishers Ltd., London, Chapter 7.
44. Andrews, E. H. (1968). *Testing of Polymers,* (Ed. W. E. Brown), Interscience, New York, Chapter 6.
45. Loveless, H. S. and McWilliams, D. E. (1970), *Polym. Eng. Sci.,* **10,** 3, 139.
46. Goldfein, S. (1968). *Testing of Polymers,* (Ed. W. E. Brown), Interscience, New York, Vol. **4,** Chapter 3.
47. Gotham, K. V. (1972). *Plastics and Polymers,* **40,** 146, 59.
48. Cessna, L. C., Jr., Levens, J. A. and Thomson, J. B. (1969). 24th Conference, RP/C Division, SPI, Paper 1-C.
49. Braden, M. (1965). *Trans. Plast. Inst.,* **33,** 103, 17.
50. Krautz, F. G. (1971). *SPE Journal,* **27,** 8, 74.
51. Anon. (1973). *Modern Plastics International,* December, 62.
52. Theberge, J. E. and Jones, R. F. (1969). *Progressive Plastics,* July, p. 31.
53. Theberge, J. E. (1970). *Modern Plastics,* March, p. 104.
54. Giltrow, J. P. (1973). *Composites,* **4,** 2, 55.
55. Perkins, C. W. and Wheeler, P. C. (1973). The Tribological Behaviour of Filled Thermoplastics, Paper presented at the Symposium on Advances in Reinforced Thermoplastics, April, Kingston-upon-Thames, Surrey, England.
56. Gotch, T. M. (October 1970). 'Engineering Applications of Reinforced Thermoplastics in Rail Transport: Carbon Fibre Reinforced Plastics Gears and Bearings.' Plastics Institute Conference on Reinforced Thermoplastics, Solihull, England, Paper 15.
57. Theberge, J. E., Arkles, B. and Cloud, P. J. (4th February and 30th September 1971). *Machine Design.*
58. Reymers, H. (1970). *Modern Plastics,* September, p. 78.
59. Tobolsky, A. V. (1960). *Properties and Structure of Polymers,* Wiley, New York.
60. Cessna, L. C., Jr. (1970). 28th ANTEC SPE, Proceedings, p. 527.
61. Moore, D. R. and Turner, S. (1974). *Plastics and Polymers,* **42,** 157, 41.
62. Caryl, C. R. (1968). *Testing of Polymers,* (Ed. W. E. Brown), Interscience, New York, Chapter 8.
63. Kambour, R. P. (1968). *Polym. Eng. and Sci.,* **8,** 4, 281
64. Titow, W. V., Braden, M., Currell, B. R. and Loneragan, R. J. (1974). *J. Appl. Polym. Sci.,* **18,** 3, 867.
65. Dunn, P. and Sansom, G. F. (1969). *J. Appl. Polym. Sci.,* **13,** 1641, 1657, 1673.
66. Vincent, P. I. and Raha, S. (June 1972). *Polymer,* **13,** 283.
67. Kambour, R. P. (1974). *Macromolecular Reviews,* (Ed. A. Peterlin), Vol. 7, Interscience, Wiley, New York and London, pp. 1–154.
68. Titow, W. V. (1974). 'Environmental Stress Cracking'. Paper presented at the Seminar on Stress Phenomena, Fulmer Research Institute, Stoke Poges, England, April.
69. Pohrt, J. (1973). 'A Critical Strain Method for the Rapid Determination of Long-term Cracking in Thermoplastics and its Dependence on Internal Structure.' One day seminar at the Polytechnic of the South Bank, London, England, June.
70. Lavengood, R. E., Nicolais, L. and Narkis, M. (1973). *J. Appl. Polym. Sci.,* **17,** 1173.

CHAPTER 7

PROCESSING OF REINFORCED THERMOPLASTICS

7.1 GENERAL FEATURES AND TECHNIQUES

Manipulation, and ultimate solidification into the desired shape, of a polymer melt constitute the cardinal feature of the two most important industrial methods of primary processing of thermoplastics: injection-moulding and extrusion. This applies also in compression-moulding and calendering. However, the former is not used commercially to a significant extent with thermoplastics unreinforced or reinforced, and the latter, though a major technique, is largely restricted to certain polymers (mainly PVC and vinyl chloride copolymers, some PTFE) and products (sheets, films and coatings). Two calendered PVC sheet materials are produced commercially in fibre-reinforced form: a type of asbestos-filled PVC flooring, and asbestos-reinforced PVC sheeting. Both have been mentioned (*cf.* Chapter 3, Section 3.2.8 and Chapter 4, Section 4.4.3).

Thermoforming, a commercially significant processing technique (secondary, in that it is applicable to thermoplastic sheet—a semi-product produced by one of the primary methods), also relies on lowering the viscosity of the material by heating. Thus the viscosity (and flow properties generally) of the polymer under the conditions of temperature and pressure used in a particular process constitute the most important single factor in thermoplastics processing. In reinforced thermoplastics melt viscosity is a function of the loading (volume fraction) and the aspect ratio of the reinforcing filler, which is a fibre in most cases. This general relationship is represented by the expression:[1]

$$\eta_r = 1 + f(L/D)^2 V_f$$

where η_r = relative viscosity (ratio of the viscosities of the filled and unfilled melts); L = fibre length; D = fibre diameter; and V_f = volume fraction of fibres.

Figure 7.1 shows the effect of fibre length on the flow distance of a glass-filled (33 wt %) nylon 6.6 in the experiments of Williams and co-workers,[1] in which the flow distance in the mould was determined

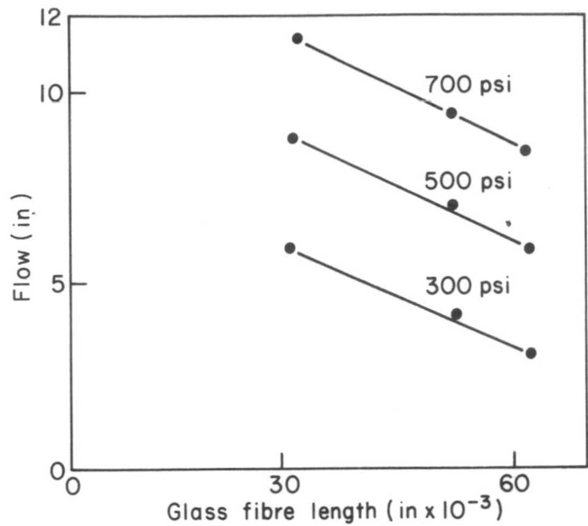

FIG. 7.1 Effect of fibre length on flow. 33% glass in nylon 6.6. (Data of Williams *et al.*:[1] reproduced with permission of the authors.)

on a one ounce, ram injection machine at the pressures shown (temperatures: cylinder 290°C, mould 100°C).

The rheological properties are also affected by the state of dispersion of the filler.

Several of the standard thermoplastic processing techniques are applicable to reinforced thermoplastics. Extrusion, compression-moulding, rotational moulding and thermoforming of sheet have all been used, but they are of minor commercial importance in comparison with injection-moulding. This technique is therefore discussed in a separate section below.

Extrusion of sections and profiles in some reinforced thermoplastic materials, for use in the automobile and furniture industries, whilst comparatively new, is now a commercial process. Partial orientation of the reinforcement fibres in the longitudinal direction can add to the strength of such products. Production of sheet by extrusion of glass-filled thermoplastic polyesters has also been introduced lately on a limited scale. More recently (in 1974) extruded sheet was described, which was experimentally produced from high density polyethylene reinforced with glass fibre and also with mica.[2] Fibre-reinforced nylon 6 sheet appeared in 1974, marketed as a development product by the Allied Chemical Corporation (see also Chapter 8).

Rotational moulding (powder casting) has been used to produce such articles as large food containers, gas and fuel tanks and machine covers in glass-reinforced polyethylene.[3,4] Large containers in unreinforced polyethylene are fairly widely used for food transport. If well designed and of good quality, and when they are properly palleted or protected in transit, they are suitable even for marine transport. However, any improvement in rigidity, such as would be conferred by glass reinforcement, is desirable providing it does not increase too much the cost of the containers. The use of other polymers (polypropylene, polystyrene and nylon) for rotational moulding on a small scale has also been reported.[3] The rotational moulding technique employed is a modification of the standard method as used for unreinforced polymers. In outline the typical procedure for a polyethylene/glass fibre composition is as follows. Polymer powder is dry-blended with glass fibre (chopped strand $\frac{1}{32}$–$\frac{1}{8}$ in, up to about 15% by weight). This gives better ultimate physical properties than the use of pulverised melt-compounded material, because the necessary comminution reduces the glass fibre length.[5] A charge is weighed out and placed in the mould which has been pre-treated with mould release. The mould is clamped shut and simultaneously rotated and tumbled in an oven for up to 10 min at about 300–400°C. It is then withdrawn from the oven, cooled (air or water spray) opened and unloaded. The process can be automated. Suppliers of commercial rotocast equipment (machinery, moulds and heating ovens) are listed in the standard reference books, e.g. the *Modern Plastics Encyclopedia.* A few process and equipment details have been published.[3,6] The main process variables and their effect on product properties were studied by Sowa[5] and Bernardo.[7]

Blow-moulding is another technique which has been applied to the production, in several reinforced thermoplastics, of such products as pipes and drums.[8]

Compression-moulding of reinforced thermoplastics is not normal industrial practice. Where it is carried out, e.g. in development or experimental situations, pre-compounded granules are commonly used. The procedure is essentially the same as for unreinforced base polymer alone, but higher temperature and occasionally higher pressure may have to be used. Some general indications of those conditions are given in Chapter 3, Section 3.3.

Thermoforming of reinforced thermoplastic sheet is another technique so far with comparatively limited commercial application. Before the advent of extruded reinforced sheet, sheet production methods involved such cumbersome operations as layer-wise build up from fibre mat and polymer film, or fusion and rolling into sheet of powdered polymer and loose fibre, or impregnation of fibre mat with polymer melt or solution followed by fusion.[9,10] Sheets so produced are expensive; even extruded sheet is not cheap and it may be anisotropic due to filler orientation, which may be further developed in forming. The major operational stages are the same as in the thermoforming of unreinforced sheet, i.e. cutting out a blank, pre-heating to soften, forming, removal from mould and finishing as necessary. However, for production runs a conventional metal-stamping press is normally used, with metal stamping dies (kirksite or aluminium), and the blanks are pre-shaped to minimise or eliminate trimming. The surface finish can be comparable to that in standard injection-moulding, but separation of resin from fibre can result in deep-draw mouldings. The main advantages claimed include comparatively short cycles and possibility of scrap-free operation. An appraisal of the technique has been published[3] as well as an account of its application to the production of an automobile lamp housing and an emission-control part in glass-reinforced polypropylene.[11]

7.2 INJECTION MOULDING OF REINFORCED THERMOPLASTICS

7.2.1 General

Essentially standard equipment and methods are used to mould reinforced thermoplastics with, however, some modifications necessitated largely by the higher viscosity and more rapid cooling of the polymer melt when it contains reinforcing fillers. The modifications are mentioned, as appropriate, in the following sections.

Whilst many reinforced thermoplastics may be moulded on ram (plunger) injection moulding machines, modern reciprocating ('in line') screw pre-plasticising machines are the most suitable and most widely used. This is because the processing advantages they offer in comparison with plunger machines are particularly important and desirable in the moulding of reinforced thermoplastics. The main advantages are better and more rapid homogenisation of the melt (including dispersion of filler in the melt when mixed feed is used), better melt temperature control, automatic metering of the amount of melt injected (the 'shot'), and more uniform mouldings. The discussion in this section relates to the modern screw-pre-plasticising equipment and process.

The well-known elements of the injection moulding process may be summarised as follows. The moulding material (here pre-compounded granules, or a mixture of polymer powder and reinforcement (*cf.* Chapter 5, Section 5.2.3.2) is fed from the hopper into the heated cylinder of the machine, where the polymer is melted ('plasticised') and the melt brought up to the required temperature by the joint effect of the external heating and the work heat generated by the action of the screw. The reinforcing filler forms a suspension in the melt, in which it should be uniformly dispersed. In some machines intermeshing twin screws are used. The rotating screw conveys the melting (and ultimately molten) material forward, building up a charge of pre-determined volume in the front of the cylinder. The charge is then injected, by controlled axial forward movement of the screw under applied hydraulic pressure, into the mould cavity via the cylinder nozzle, channels in the mould (sprue and runners) and the gate (*cf.* Fig. 7.2). The mould ('tool') is at a temperature lower than the melting temperature of the polymer, so that the melt solidifies, the resulting moulding ('part') retaining permanently the shape of the cavity. Tools are designed to open along a convenient plane ('split line') allowing the moulding to be ejected by ejector pins.

The injection-moulding process offers high speed production at relatively low cost. It can produce many hundreds of thousands of components from one injection mould, and operate automatically or semi-automatically. The components produced can vary widely in weight and size, from small mouldings weighing a few ounces to large ones, tens of pounds in weight.

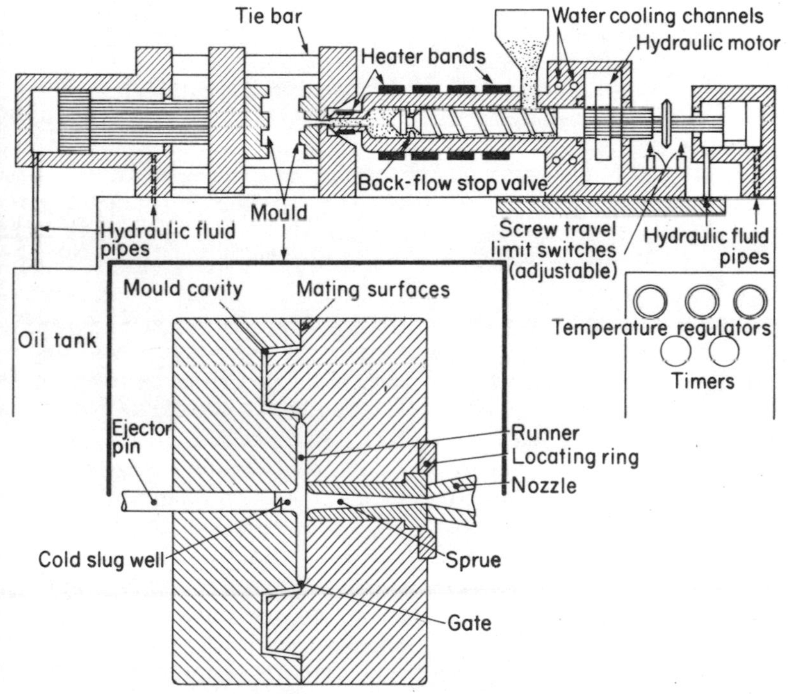

FIG. 7.2 Screw preplasticising injection moulding machine and mould, schematic representation. (Reproduced, with permission, from TS Note G103, ICI Plastics Division.)

Traditionally, injection machine capacity has been rated according to the shot weight, i.e. weight of plastics material which can be injected in one stroke. More recently rating by locking (clamping) force has been quite common: this is the force, usually expressed in tons (locking tonnage) that can be applied to hold the tool halves closed against the pressure of the material being injected. Thus a small machine may have a shot weight of 60 g ($2\frac{1}{2}$ oz) and a locking force of 45 tons, and a large machine a shot weight of 50 lb and a locking force of 2000 tons.

7.2.2 Equipment

7.2.2.1 Injection Moulding Machine

A single-screw injection moulding machine with a simple mould is shown schematically in Fig. 7.2. The principal points to be considered in selecting a machine (not specifically for processing reinforced thermoplastics) have been reviewed by Taylor.[12]

The machine features desirable or necessary in the moulding of reinforced (especially fibre-reinforced) thermoplastics are summarised below.

7.2.2.1.1 Injection pressure and speed

The machine should be capable of high injection pressures, up to and over 20 000 psi. In general, the fastest possible injection rate is desirable (normally slower with PVC and fluoropolymers), but an injection-speed control valve should be fitted. The surface finish of reinforced thermoplastic mouldings can be strongly affected by speed of injection, *inter alia*, normally improving with increasing speed.

7.2.2.1.2 The hopper

This should preferably be of the hopper dryer type, to counteract moisture pick-up by the moulding compound, particularly where that had been pre-dried (*cf.* Table 7.1) and is transferred directly from the drying oven into the hopper. With reinforced polycarbonate it is especially important to ensure a very low moisture content (less than 0·02% by weight[13,14]) because the polymer is prone to thermal degradation, accompanied by evolution of CO_2, at moulding temperatures in the presence of water, and the physical properties of the mouldings deteriorate. Streaks and—in extreme cases—bubbles can also appear. Bubble formation under standard heating conditions is the basis of a practical test recommended by Bayer for detection of unacceptably high moisture content levels in polycarbonate.[15] A dehumidifying hopper dryer is therefore desirable for polycarbonate to prevent any reintroduction of moisture from the hot air (which may contain it in considerable concentration if an ordinary hopper dryer is used).

7.2.2.1.3 THE CYLINDER

Nitrided steel cylinders are acceptable, although bimetallic cylinders, e.g. with Xaloy* liners, give better service.[3,12,16] Three heating zones (electric band heaters) are normally sufficient, but up to five can be available on some machines. Temperature control should be accurate for uniform melting and prevention of overheating of the polymer, the controlling thermocouples preferably mounted in slots under the band heaters ('shallow' control system†). The cylinder nozzle should also be heated (separately controlled heater); it should be short, with a straight-through bore as large as is usable with the sprue bush employed on the tool. For fairly large components this would typically be $\frac{1}{4}$ in. It is inadvisable to have a sprue bush with a diameter of less than $\frac{1}{8}$ in. A reverse taper of about 2 or 3° can be usefully employed to ensure that the sprue comes away cleanly from the nozzle. Except where very high melt temperatures must be used (resulting in low melt viscosities) it is not normally necessary to employ a shut-off nozzle with reinforced thermoplastic materials. Such nozzles may occasionally be used with reinforced nylons, although the tendency for the melt to drool from the nozzle is much less when reinforcement is present.

7.2.2.1.4 THE SCREW

The recommended surface finishes (in order of increasing durability) are: nitrided flights, Stellite‡ flights, chrome plating overall.[16] A general-purpose screw design is satisfactory for most reinforced materials, but for fibre-reinforced PVC a low-shear screw, with low compression ratio, is recommended.

7.2.2.1.5 LOCKING MECHANISM

The general recommendation is that this should be capable of providing a locking force of 5–10 tons in^{-2} projected area of the moulding.[16,17]

7.2.2.2 THE MOULD

7.2.2.2.1 MATERIALS

For long production runs hardened tool steel is the preferred material, to resist wear of the surfaces of the mould cavity, cores, sprues, runners and gates. The area around the gate on the surface of cores is subjected to the greatest wear: wear can also be caused by material flashing across the end of a core. However, general wear is not excessive when pre-compounded moulding stock is used (see also Section 7.2.3.1 below).

* Xaloy Inc., USA.

† This control system, originated by ICI Plastics Division, minimises temperature fluctuation ('hunting').

‡ Stellite Division, Union Carbide Corp., USA.

Hardenable stainless steel tools have been used to good effect. Apart from the question of surface hardness and physical resistance, their corrosion resistance is also advantageous in resisting rusting in storage and in operation with materials whose decomposition products may be corrosive, e.g. PVC, some fluoropolymers. Whilst, e.g. DuPont's Tefzel ETFE copolymer can be quite successfully moulded in tool-steel cavities, DuPont point out that any chrome or nickel plating must be of high quality because the strong adhesion between the polymer and the plating can cause stripping.[18]

For prototype tools aluminium is commonly used, although epoxy resin moulds are also sometimes employed.

7.2.2.2.2 TOOL DESIGN

The general principles of good tool design apply equally to moulds for reinforced thermoplastics. Among these, Gray[19] singles out balanced runner systems for equalised mould filling rates and the use of radii on external corners and fillets on internal corners to minimise stress concentration in the mouldings. Two general features which are somewhat modified by the requirements of successful moulding of reinforced (especially glass-fibre-reinforced) thermoplastics are venting and ejector pins. By and large more and larger vents will be required than with corresponding unreinforced polymers to counteract tendency to entrapment of hot gases enhanced by the faster moulding rates and higher temperatures used with reinforced materials. Proper venting facilitates rapid filling and improves weld strength. Larger vents can be tolerated because reinforced melts are more viscous and set up more rapidly (since volume for volume the shot contains less molten polymer than unreinforced material): both factors reduce flashing. More ejector pins will normally be needed than with unreinforced mouldings, especially when producing parts with thin sections and/or large areas: stripper plates are desirable in many cases. Proper positioning of ejector pins is also particularly important.

Mould design points of special interest in the moulding of reinforced thermoplastics include:[16,17,20–22]

(i) Sprues: The sprue should be as short as possible, highly polished, tapered, with orifice diameter similar to (marginally larger than) that of the cylinder nozzle. Cold-slug wells are a desirable feature.

(ii) Runners: These should normally be of large cross-section (full round or trapezoidal), and—like the sprues—short and well polished. These factors work to minimise pressure losses due to cooling and friction. Hot runner moulds are sometimes recommended for use with reinforced thermoplastics.[21,23]

(iii) Gating: In principle, all the common types of gate (sprue, side, pinpoint, tab, ring, fan, submarine) may be used with reinforced thermoplastics, and the gating may be single or multiple. In practice the choice is decisively influenced by the nature of the polymer (and in particular the melt viscosity), the filler (with fibres, fibre length and content are important factors) and the moulding. As far as it is possible to generalise, the following points may be noted. The position and size of the gate are even more important in obtaining good mouldings with reinforced thermoplastics than with the base polymer alone. This is because of the generally greater viscosity of the former, their quicker setting-up, and the possibility of orientation of the fibres (with the attendant possible differential shrinkage and hence warping problems). Central, symmetrical and balanced gating is important to avoid warping. Multiple gating may be used for the same reason, but it is not otherwise desirable, as the resulting weld lines can be a problem. Wherever possible the gates should be full, round or rectangular, located at (or as close as possible to) the thickest part section, to avoid formation of sinks and voids. The size of such gates should be between two-thirds and the full width of the section thickness. Gate lands should be short, preferably 0·02–0·04 in. With most fibre-reinforced materials pinpoint gating is not favoured; *inter alia* it can promote fibre orientation. In general round gates should not be less than 0·04 in (down to about 0·02 in with fibre-filled nylon 6 or nylon 6.6).

Correct mould temperature control is particularly important in moulding reinforced thermoplastics. Quality of surface finish, stress level, shrinkage, tendency to warping, appearance of weld lines, are all influenced by the temperature of the mould which, with many reinforced materials may have to be accurately maintained above 100°C. In good modern practice the heating is commonly effected by circulating oil through channels in the tool (see also next section). The dimensions and distribution of the channels, as well as their distance from the cavity surface are important considerations in mould construction.[21] Occasionally moulds are heated by electrical cartridge heaters, fitted to the mould halves. In either case separate control of the temperature of the two halves of the tool can be important. For example, warping of some mouldings may be counteracted by keeping the mould halves at different temperatures; also, where long cores are employed, it is common to run these at lower temperature than the cavity.

7.2.2.2.3 PART DESIGN

The inter-relationship between the design of a mould and that of the mouldings it is intended to produce is largely self-evident. Some features of the relationship were also indicated in the preceding sub-section.

Quite apart from this interdependence, the design of a moulding will be influenced by the relevant mechanical and thermal properties of the moulding material. Reinforced thermoplastics differ considerably from their base polymers in several of the properties important in this context. The main part design points to which these differences give rise are briefly outlined below.

The greater strength and stiffness of reinforced thermoplastics make it possible to use thinner sections to meet a given specification. Depending on the material, thickness (and hence material amount), reductions of up to 50% may be possible in comparison with unreinforced material.[3] However, uniformity of section thickness is important: changes in thickness—especially if the transition is abrupt—can cause considerable distortion in a moulding.[24,25] This can be aggravated if the resulting flow pattern gives rise to orientation of reinforcing fibres, with attendant differential shrinkage effects. In general, thin sections can be produced, but tendency to warping is an ever-present potential difficulty with such sections in fibre-reinforced material. Careful part design should be combined with control of the relevant process factors (*cf.* Section 7.2.3) to counteract this potential fault. Moulding shrinkage of reinforced thermoplastics is much lower than that of the base polymers alone (*cf.*, e.g. Chapter 3, Section 3.3). This is an important factor in part and mould design. Moreover, where the reinforcement is fibrous, orientation of the fibres can produce pronounced anisotropic shrinkage effects (*cf.* Fig. 7.3) and/or modify any tendency to differential shrinkage arising through orientation of the molecules of the base polymer. These points must be considered in part design. Prototype parts should be produced wherever possible, to determine the amount and pattern of shrinkage: this is

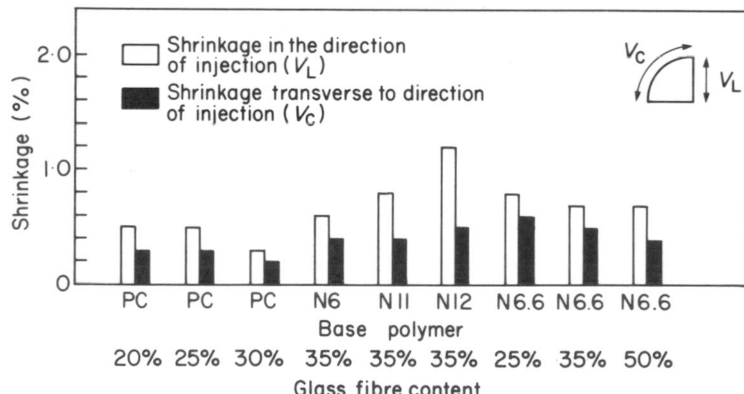

FIG. 7.3 Processing shrinkage of glass-fibre reinforced thermoplastics, determined on quarter-disc mouldings 2·9 mm thick. PC: polycarbonate; N: nylon. (Data of Lucius;[21] reproduced with permission of Carl Hanser Verlag.)

particularly important where the parts are complex and/or high tolerances are required.[17]

The combined effects of lower shrinkage and greater stiffness of reinforced thermoplastic mouldings can make removal of the part from the mould difficult. It is usual to allow adequate taper (0·5–3°) on all surfaces in the direction of ejection of the moulding to facilitate ejection.[16,17,20–22] Undercuts are avoided for the same reason. However, it has been claimed[3] that parts with little or no draft in the side-walls and other segments can be successfully produced if the tool contains a sufficient number of ejector pins suitably placed.

Recommendations have been published[3] for the design, proportions and dimensions of fillets and radii, ribs, studs, bosses and screw holes in reinforced thermoplastic parts.

7.2.2.3 Ancillary Equipment

Essentially the same ancillary equipment is used with reinforced thermoplastics as with the base polymers alone, but certain points assume particular importance. These are mentioned below.

7.2.2.3.1 DRYING EQUIPMENT

Conventional tray dryers, vacuum dryers, de-humidifying dryers and the already mentioned hopper dryers can be used. Some reinforced materials do not have to be pre-dried if used directly out of sealed containers in which they are delivered. With those materials which require pre-drying (*cf.* Table 7.1) de-humidifying equipment is preferable and in particular a dehumidifying hopper dryer is desirable.

7.2.2.3.2 MOULD HEATING UNIT

This is another piece of ancillary equipment whose use and quality take on particular importance when injection moulding reinforced thermoplastics. As has been mentioned, mould temperatures above 100°C may be necessary as well as separate control of tool parts at different temperatures. Thus the older type of water heater still common in many moulding shops is not generally suitable: either oil circulating heaters or electrical cartridge heaters should be used. Two typical good quality modern mould heating units are shown in Figs. 7.4 and 7.5.

The ratings quoted in the captions of Figs. 7.4 and 7.5 are approximate, applicable as general guidance figures, e.g. with glass-reinforced nylon and PPO. Specific applicability is determined by various factors, including the weight and construction of the mould, as well as nature of the reinforced material.

7.2.2.3.3 GRINDER

Re-grind can be handled in the usual way, making use of either a small

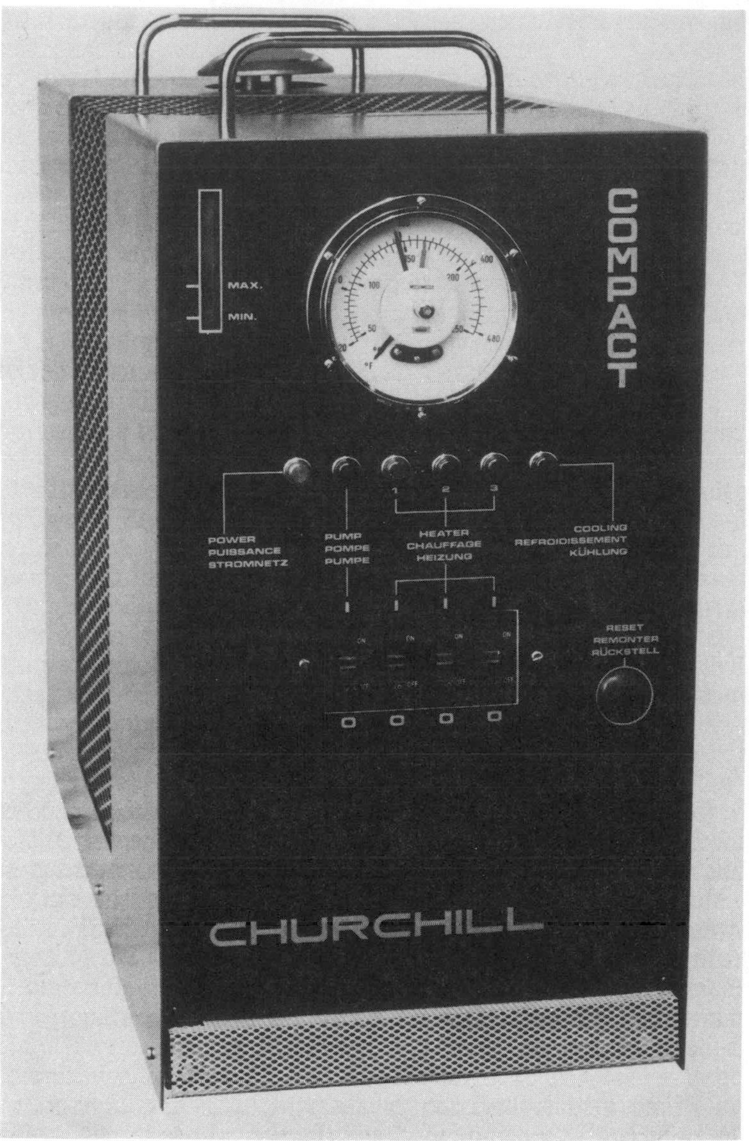

FIG. 7.4 5 kW oil-circulating temperature controller (Churchill Compact). Operating range 50–200°C, for use to approximately 6 oz shot weight. (Courtesy of the Churchill Instrument Co. Ltd.)

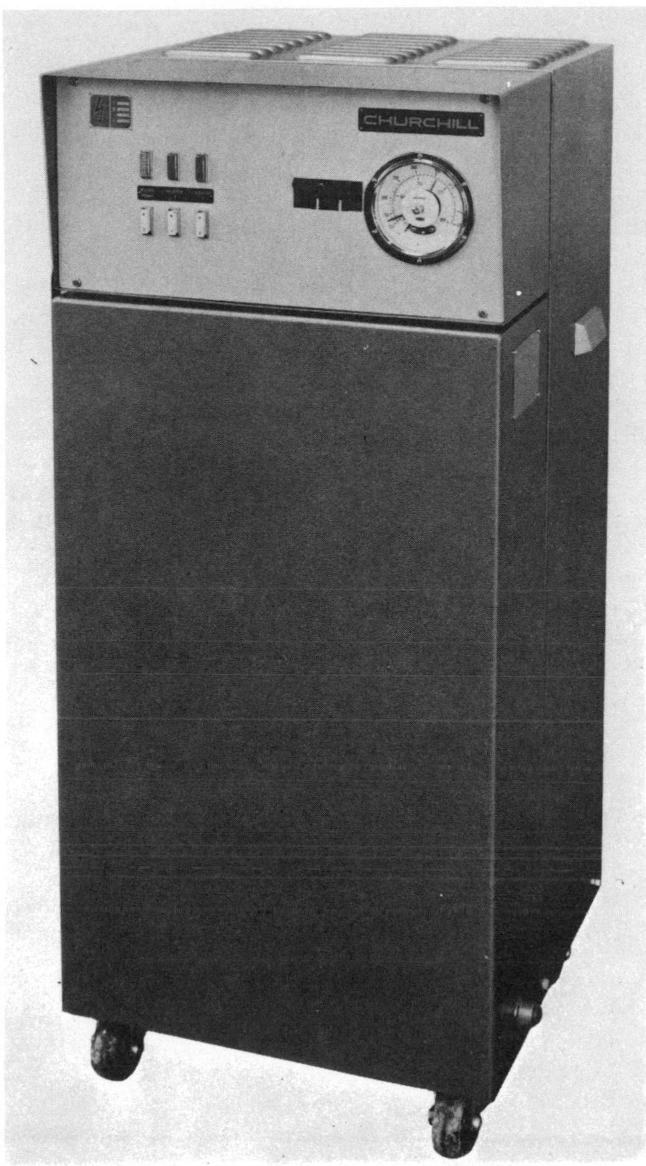

FIG. 7.5 13 or 20 kW oil-circulating temperature controller (Mark II 9). Operating range 50–200°C for use with shot weight capacities of up to 30 oz. (Courtesy of the Churchill Instrument Co. Ltd.)

grinder positioned close to the injection machine, or a larger centralised installation. The use of re-grind is discussed in Section 7.2.3.3 below.

7.2.3 Moulding Process

7.2.3.1 General Points

In normal industrial moulding of reinforced thermoplastics the actual shot size used is 50–75% of the rated, nominal capacity of the machine. Because of the generally higher melt viscosities in comparison with those of the base resins alone, injection pressures are higher, by up to about 80% in some cases. Reinforced materials are normally injected at the highest possible speed: this gives rapid filling of the mould and hence promotes good surface finish, weld line integrity and reduces orientation of fibrous reinforcement in the moulding. However, high injection speeds may occasionally result in flashing in the tool and the speed may have to be adjusted downwards until flash-free mouldings are obtained.

Shorter cycle times are possible with reinforced thermoplastics than with the base polymers alone (for comparable section thickness, gating, and relevant processing factors), because the set-up (solidification in the mould) is more rapid (*cf.* Section 7.2.2.2.2, 'Tool design'), and also because of the greater stiffness and higher temperature for deformation, both of which make it possible to remove warm parts from the mould without damage.

Reference has already been made to the fact that screw speed and back pressure (the hydraulic pressure applied to resist backward movement of the screw during pre-plasticisation) influence the average length of reinforcing fibres in injection moulding, and hence affect the mechanical properties of the mouldings (*cf.* Chapter 5, Section 5.3). Filbert[26] has shown that the effect of screw speed in this respect is somewhat greater than that of the back pressure (see Section 7.2.3.3 below).

To maintain the fibre length in the finished component the screw speed and back pressure should be kept to a practicable minimum, consistent with obtaining a homogeneous melt. Typically, a minimum back pressure of 50 psi can give good mixing without significant fibre degradation. With glass-reinforced nylon 6.6, zero back pressure is recommended, or the minimum necessary to prevent the screw 'augering', i.e. failing to pick up resin.[16] The recommended range of screw speed is 30–60 rpm.[17,19] In practice about 50 rpm is satisfactory for most materials.

The measurement of the length of flow in a spiral-channel mould is a common method of assessing processing properties of injection-moulding materials. The spiral flow length is principally governed by the melt viscosity and the setting up rate (which, *inter alia*, determines the rate of freezing in the gates and runners). Maxwell[27] points out that addition of glass fibre to nylon 6.6 makes the setting up more rapid and sharper due largely to the reduction of the latent heat of fusion: the magnitude of the resultant reduction in spiral melt flow is also a function of the melt (cylinder) temperature varying between about 40% at low temperatures and 15% at high temperatures. In general, the effect varies with the nature of the base polymer. It is least in glass-fibre-reinforced polycarbonate (spiral flow reduction about one-half that of nylon).

In normal moulding situations the main factor influencing the flow pattern in the mould, and the associated fibre orientation effects which can arise in moulded reinforced parts, is the geometry of the mould cavity.[28,29] The rheological properties of the melt, and mould temperature, also play a part. However, recent work by Schmidt[29] indicates that up to moderate loading levels neither glass fibres nor spheres critically affect the flow pattern in the mould. Similarly, the effect of mould temperature on flow pattern assumes substantial significance only as the temperature of the mould approaches that of the melt. The flow pattern arising under a given set of conditions (of mould geometry and temperature and rheology of the melt) affects the orientation of fibrous fillers. Anisotropy caused by fibre orientation is a factor in differential shrinkage, and hence in the associated tendency to warping of thin-section parts moulded in glass-fibre-reinforced materials. This cannot always be effectively remedied by the common expedients of running the mould halves at different temperatures and ensuring symmetrical gating (*cf.* Section 7.2.2.2). Fibre orientation in a reinforced thermoplastic part may be detected and/or assessed by comparatively simple methods, including:

(a) microscopic examination of thin parts, or prepared sections, or fibres near the surface of a moulding;
(b) inspection of the disposition of fibres, after careful removal of the base polymer from a moulding or a prepared section, by simple dissolution (where suitable solvents are available), dissolution by degradation, e.g. with strong acids—*cf.* Chapter 4, Fig. 4.14, where the polymer is not readily soluble, or 'burn-out' (degradation and removal by controlled heating—*cf.*, e.g. Chapter 4, Fig. 4.2).

Application of some variants of these methods to the determination of fibre orientation in several base polymers has been described by Fucella.[28]

7.2.3.1.1 WEAR OF MACHINE AND TOOL SURFACES

Passing reference has already been made to this subject in connection with materials for machine and mould surfaces (Sections 7.2.2.1 and 7.2.2.2 above). In general, the tendency to increased mould and machine wear by reinforced (especially glass-fibre-reinforced) thermoplastics in comparison with the base resins alone is less than might perhaps be expected. When proper moulding conditions are observed, wear caused e.g. by glass-fibre-filled nylon 6 or 6.6 or polyethylene terephthalate, can be negligible.[16,30,31] It has been claimed[31] that, where it occurs in the processing of reinforced thermoplastics, wear of cylinder and screw surfaces may be due to corrosion, rather than to abrasion or erosion by the filler. When pre-compounded moulding stock is used, with the reinforcement well dispersed in the base polymer, the resulting effective 'encapsulation' of the filler fibres is certainly a factor in minimising abrasion effects.

7.2.3.2 Temperatures and Associated Conditions

Processing conditions for a number of thermoplastics containing various reinforcements are summarised in Table 7.1. The temperatures quoted in the table (in particular the melt and cylinder temperatures) are intended as indications of typical conditions. The ranges given in the 'cylinder temperature' column encompass the temperatures of the separate heating zones (commonly three) and the nozzle. The individual zone temperatures are normally set to increase from the back (hopper) end of the cylinder towards the front. The temperature of the rear zone in particular has been found (by Filbert[26]) to be a factor in the effect of processing on average length of fibrous reinforcement and hence the properties of reinforced thermoplastic mouldings. Filbert's data are shown in Table 7.2.

Reinforced thermoplastic mouldings, like unreinforced ones, may be annealed by heating to relieve moulded-in stresses. The heating medium may be either air or oil: oil is more effective and quicker. Non-acidic, refined mineral oils or silicone oils may be used, of which several commercial versions are available.* It is self-evident that the oil should not promote stress cracking of the base polymer. The annealing conditions will also depend on the material. It is usual for the moulder to consult his material supplier when deciding on the annealing treatment in a particular case. As an example, a nylon 6.6 moulding may be annealed at 180°C, nylon 12 at 140°C, polycarbonate at 120°C and acetal polymer at 150°C. The annealing time is related to section thickness and may range, according to the material, from about 5 min mm^{-1} upwards. The cooling after the treatment should be comparatively gradual (say, 20–40°C h^{-1}).

7.2.3.3 Reprocessing (Use of Re-Grind)

In line with the practice generally followed with unreinforced thermoplastics, sprues, runners and clean rejects in reinforced material are normally re-ground and used up by adding to virgin moulding stock.

As has been mentioned in Chapter 5, Section 5.3, re-working a reinforced thermoplastic may *a priori* be expected to influence the

* For example from Houghton & Co. (102 oil) or Gulf Oil Corp. (Security oil 205) in the USA; ICI Nobel Division (silicone oil DP 190), Shell Chemical Co. (Risella oil 33, Rotella IOW/30), Union Carbide Ltd. Chemicals Division (Ucon 50/HBZ/280/X) in the UK.

TABLE 7.1

Injection-Moulding of Reinforced Thermoplastics: Typical Processing Temperatures

Polymer and reinforcement	Pre-drying (air oven) Temperature (°C)	Pre-drying (air oven) Min. time (h)	Cylinder temperature (°C)	Melt temperature (°C)	Mould temperature (°C)	Remarks
Nylon 6/glass fibre	110[a]	2	245–290	260–290	80–120	De-humidifying or vacuum ovens should be used for drying nylon moulding compounds
Nylon 6.6/glass fibre	110[a]	2	265–300	270–300	80–140	
Nylon 6.6/glass beads	80–100[a]	2	260–305	270–300	20–90	
Nylon 6.6/asbestos fibre	80–100[a]	2	260–305	250–280	65–140	
Nylon 6.6/Fybex	90–100[a]	2	280–300	up to 300	105	
Nylon 6.12/glass fibre	90–100[a]	2	250–290	260–290	80–120	
Nylon 12/glass fibre	80[a]	4	180–300	220–270	40–120	
Nylon 12/glass beads	80[a]	4	240–270	250–270	60–100	
Acetal/glass fibre	Normally not necessary (80)		180–215	180–210	65–120	Melt temperature about 250°C results in serious decomposition
HD Polyethylene/glass fibre	Normally not necessary		250–310	220–280	30–45	
Polypropylene/glass fibre	Normally not necessary		180–290	240–270	30–90	
Polypropylene/asbestos fibre or talc	Normally not necessary (100–110)		230–280	260–270	50–90	
Polypropylene/Fybex	Normally not necessary (80–90)		230–270	245	80	

Table 7.1—*continued*

Polymer and reinforcement	Pre-drying (air oven) Temperature (°C)	Pre-drying (air oven) Min. time (h)	Cylinder temperature (°C)	Melt temperature (°C)	Mould temperature (°C)	Remarks
Polyester (PET)/glass fibre	120–130[a]	4	260–290 (decomposition above 300)	240–290	120–140 (or cold for amorphous moulding)	For Arnite A340, AKZO recommend highest zone temperature on the hopper side of the cylinder
Polyester (PBT)/glass fibre	120–130[a]	4	230–250	230–250 (deterioration above 270)	40–120	
Polycarbonate/glass fibre	120	24	270–340	270–320	80–120	Moisture content of granules must be <0·1%
Polysulphone/glass fibre	90[a]	20	340–400	300–400	70–150	
Polysulphone/Fybex	95[a]	About 18	340–370	350–380	150	
Modified PPO/glass fibre	105–115	2–4	275–315	250–300	80–110	
Modified PPO/Fybex	95–105	About 4	310–325	320	105	
Polystyrene/glass fibre	70–80	4	230–310	240–290	60–80	Pre-drying recommended for good surface finish and strength
Polystyrene/asbestos fibre	70–80[a]	4	200–240	240	20–50	

Table 7.1—*continued*

Polymer and reinforcement	Pre-drying (air oven) Temperature (°C)	Pre-drying (air oven) Min. time (h)	Cylinder temperature (°C)	Melt temperature (°C)	Mould temperature (°C)	Remarks
SAN/glass fibre	70–80[a]	4	260–290	270–280	70–80	Cylinder temperatures appropriate to large components (long flow required)
ABS/glass fibre	80–90	8	260–290	270–280	70–90	
ABS/Fybex	80–90[a]	About 8	250–275	250–270	95	Moulding temperature of about 295°C recommended for moulding for electroplating
ETFE/glass fibre	Normally not necessary		275–345	300–340	20–175	Special corrosion-resistant working surfaces recommended
PVC/glass fibre	80	8	150–195	190 (max.)	25	
Polyurethane/glass fibre	90–105	2	165–210	180–200	40–65	

[a] May be moulded in the 'as received' state (from sealed packages), but if exposed to atmosphere for appreciable time should be dried in conditions indicated, preferably in a de-humidifier dryer.

TABLE 7.2
Average Fibre Length and Mechanical Properties of Moulded Bars as a Function of the Temperature in the Rear Zone of the Cylinder
(Reprinted, with premission, from *SPE Journal* **25**, 1, Jan. 1969)

Cylinder zone temp. (°F)			Screw retraction (s)	Avg. fibre length (mil)	Tensile strength (psi)	Flexural modulus (psi)	Izod (ft lb in^{-1})
Rear	Cent.	Front					
550	525	525	16·5	21·9	26 800	1 415 000	2·4
525	525	525	16·8	21·4	26 620	1 403 000	2·4
500	525	525	18·9	20·2	26 400	1 376 000	2·3
475	525	525	20·5	19·7	26 250	1 369 000	2·2
450	525	525	21·1	19·5	26 070	1 356 000	2·1

Notes:
1. Tensile strength—ASTM D-638.
   Flexural modulus—ASTM D-790, average of five specimens.
   Izod, notched—ASTM D-256.
2. 6 oz reciprocating screw machine.
3. Conditions: 48 rpm, 0 back-pressure, 60 s overall cycle.

properties (especially mechanical properties) either through reduction of the length of the fibres (which will be the reinforcing filler in most cases) or some heat degradation of the base polymer, or both. In practice the effects vary with the material and the processing conditions. If the latter are properly controlled polymer degradation is not normally a serious problem. For instance ETFE copolymer (Tefzel) can be re-moulded several times without excessive deterioration of properties, although the colour may change considerably. Some typical effects of re-cycling, coupled with excessive hold-up time for unreinforced Tefzel are shown in Table 7.3. Precautions to be observed with this material include prevention of contamination (which can promote decomposition) and use of corrosion-resistant, clean machinery.

TABLE 7.3
Some Effects of Re-Cycling on Tefzel[R] *ETFE Polymer (60 min Holdup Time)*
(Reprinted, with permission, from 'Technical Bulletin PIB 7, Du Pont de Nemours International S.A.)

No. of cycles	% Change in melt flow no.	Colour	Tensile properties at 73°F (23°C)	
			Tensile (psi)	Elongation (%)
1	40	Creamy	6 500	150
2	90	Tan	6 100	150
3	120	Dark brown	6 000	150

By and large, reduction of the length of reinforcing fibre is the most important single cause of changes in the properties of re-grind, and hence of mouldings which incorporate it in significant proportions. However, a lot depends on the actual material and the processing conditions: in some reinforced thermoplastics the effects of re-cycling can be negligible. For example, glass-fibre-reinforced modified PPO (Noryl) has been re-ground and re-processed repeatedly without substantial change in mechanical properties as illustrated by the data of Table 7.4.

With glass-fibre-reinforced nylon 6.6 Filbert[26] found that the fibre length and length distribution—and hence the mechanical properties—were reduced by re-cycling; the screw speed and—as already mentioned, *cf.* Table 7.2—the temperature of the rear zone of the cylinder on the injection-moulding machine were particularly important factors (*cf.* Fig. 7.6 and Table 7.5).

Some effects of re-cycling on Fybex-reinforced nylon 6.6 are shown in Table 4.11, Chapter 4.

For the above reasons the percentage of clean, good re-grind which may safely be added to virgin material in commercial moulding operations will vary considerably, depending on the material and the conditions. As

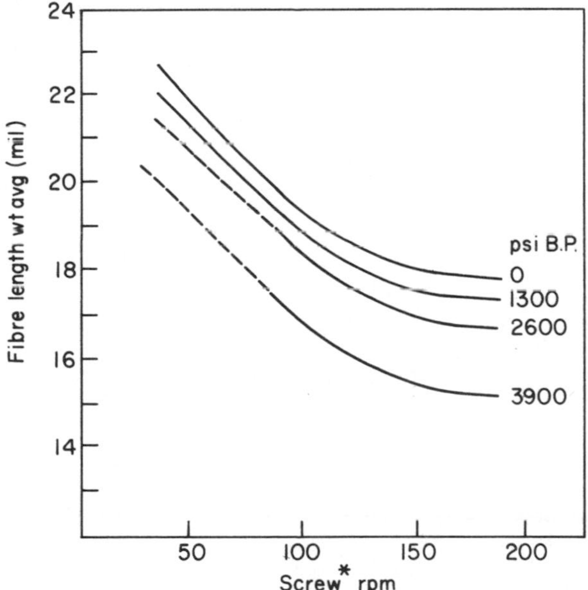

FIG. 7.6 Fibre length *vs* screw speed at indicated back-pressures for 33% glass-reinforced nylon. *General purpose—40 mm diameter. (Data of Filbert: reprinted, with permission, from *SPE Journal*, **25**, 1, January 1969.)

TABLE 7.4

Physical Properties of Noryl^R and Glass Reinforced Noryl^R after Repeated 100% Re-Grinding

(Reproduced, with permission, from Technical Bulletin NE-1M, 12–71, Engineering Polymers Ltd.)

Property	Virgin Noryl^R	100% Re-grind		
		1st Cycle	3rd Cycle	5th Cycle
Tensile yield strength kgf cm^{-2} (psi)	650 (9 215)	640 (9 070)	670 (9 500)	680 (9 640)
Tensile modulus kgf cm^{-2} (psi)	25 000 (355 000)	25 000 (355 000)	25 000 (355 000)	25 000 (355 000)
Elongation (%)	60	60	60	60
Impact, notched ($\frac{1}{4}$ in)	4	4	4	4

Property	Virgin Noryl^R GFN2 (20% Glass fibre)	100% Re-grind	
		1st Cycle	3rd Cycle
Tensile yield strength kgf cm^{-2} (psi)	1 020 (14 600)	1 025 (14 700)	1 025 (14 700)
Tensile modulus kgf cm^{-2} (psi)	65 000 (930 000)	65 000 (930 000)	65 000 (930 000)
Elongation (%)	4–6	6	5
Impact, notched ($\frac{1}{4}$ in)	1·6	1·6	1·6

TABLE 7.5

Re-Moulding[a] Glass-Reinforced Nylon (33% Glass Fibre by Weight): Effect on Fibre Length and Some Physical Properties
(Data of Filbert—reprinted from *SPE Journal*, **25**, 1, Jan. 1969)

	Avg. fibre length (mil)	Loss (% of virgin)	(D-638) Tensile strength (psi g, 0·2 in min^{-1})	Elongation (%)	(D-790) Flexural modulus (psi)	(D-256) Izod (ft lb in^{-1})	(D-1822) Tensile impact (ft lb in^{-2}) Type S	(D-732) Shear strength (psi)	Mould shrinkage (in in^{-1})
Virgin resin	31·4	—							
1st Moulding	21·7	30							
dry-as-moulded			25 900	2·6	1 403 000	2·4	39·7	11 990	0·000 9
Saturated at 100% RH			9 670	3·0	503 000	—	—	—	
2nd Moulding (100% rework)	19·5	38							
dry-as-moulded			24 750	2·7	1 320 000	2·1	30·9	11 490	0·001 2
Saturated at 100% RH			9 040	3·3	498 000	—	—	—	
3rd Moulding (100% rework)	17·0	46							
dry-as-moulded			23 000	2·7	1 287 000	1·8	27·8	11 460	0·001 7
Saturated at 100% RH			8 270	3·5	434 000	—	—	—	
4th Moulding (100% rework)	14·7	53							
dry-as-moulded			21 640	2·8	1 188 000	1·4	21·8	11 300	0·002 1
Saturated at 100% RH			7 980	4·1	426 000	—	—	—	
Reground (4X) resin	13·6	57							

[a] Least severe processing conditions were used in re-grinding and re-moulding, including sharp knives and largest mesh screen in granulator and slow screw speed with zero back pressure in the injection machine.

TABLE 7.6
Nature, Causes and Remedies of Moulding Faults

Fault	Likely cause	Suggested remedy[a]		
		Machine		Mould
Short mouldings	Insufficient material Inadequate flow Melt cooling too fast	Increase feed Increase injection pressure Increase injection speed Increase melt temperature Increase mould temperature Enlarge size of nozzle Use larger machine	and/or	Vents, sprue and runners, gates
Part undersize	Insufficient material or excessive shrinkage	Increase injection speed Increase injection pressure Increase screw forward time Increase cylinder temperature		Adjust mould temperature[b] Enlarge runners and gates
Part oversize	Over-filling; shrinkage lower than allowed for	Reduce cylinder temperature Increase clamping pressure Shorten overall cycle Reduce injection pressure		Increase mould temperature
Shot (part weight) variation	Intermittent heat variation in feed and/or melt temperature and/or pressure: bridging in hopper	Check and adjust as necessary: stock feed, e.g. for bridging in hopper heater bands all temperature controls and thermocouple contacts machine hydraulics (for pressure fluctuation)		

Table 7.6—*continued*

Fault	Likely cause	Suggested remedy[a]	
		Machine	Mould
Sink marks	Insufficient material in mould in relation to actual shrinkage on cooling	Increase screw forward (dwell) time Increase injection pressure Increase injection speed Make sure moulding stock is properly dried Use larger machine	Adjust mould temperature Enlarge runners Reduce gate lands
Weak weld lines	Too rapid cooling of melt entering mould and incomplete fusion on contact of divided melt streams	Increase injection speed Increase injection pressure Increase melt temperature Re-dry material	Increase mould temperature Enlarge vents Re-locate or re-design gates
Burn marks (charred areas)	Surface degradation through overheating in the presence of trapped hot gases or air in mould	Reduce melt temperature Reduce injection speed Enlarge nozzle opening	Enlarge vents (size and/or number) Enlarge and/or re-locate gates
Degradation (discoloration and/or decomposition)	Overheating: too-high melt temperature or excessive dwell time in cylinder	Reduce dwell time (by reducing overall cycle), purge cylinder Reduce melt (cylinder) temperature Check all temperature controls Check nozzle and cylinder for hold-up points	

Table 7.6—*continued*

Fault	Likely cause	Suggested remedy[a]	
		Machine	Mould
Splash marks ('mica marks', 'silver streaks')	Presence of moisture (left in material by insufficient drying or, less commonly, formed by decomposition)	Re-dry material (*cf.* Table 7.1): if this is not effective, reduce melt temperature (to minimise decomposition) and/or	enlarge gates (to reduce frictional heating)
Poor surface finish	(i) Poor flow of melt	(i) Increase melt temperature Increase injection pressure and speed Increase shot size	(i) Increase mould temperature
	(ii) Poor finish on cavity surface	(ii) —	(ii) Polish cavity surface
Flashing	Leakage of melt at parting lines in mould because of insufficient locking and/or too low melt viscosity	Reduce melt (cylinder) temperature Reduce injection speed Reduce injection pressure Increase clamping pressure	Ensure proper mould alignment; check mating surfaces, dowel pins and ejector system. Ensure mould locks before injection starts
Non-uniform colour dispersion	Incomplete homogenisation of melt in cylinder	Increase back pressure Increase screw speed Reduce cylinder temperature (to promote higher shear in mixing)	Increase mould temperature (to counteract possible effect on surface finish of the suggested machine adjustments)

Table 7.6—*continued*

Fault	Likely cause	Suggested remedy[a]	
		Machine	Mould
Warping distortion	(i) Differential shrinkage effects (especially in thin, fibre-filled parts)	(i) Increase injection pressure Increase screw-forward time	(i) Check that part and mould design are suitable. Check, balance and (possibly) reduce mould surface temperature (differential heating of mould halves may be beneficial)
	(ii) Distortion during ejection (less common with reinforced thermoplastics because of higher distortion temperature and greater stiffness)	(ii) Increase mould cooling time	(ii) Ensure uniformity of part ejection and check handling after ejection
Voids and bubbles	Essentially as for sink marks		

[a] More than one remedy may have to be applied. All may have to be tried before final adjustments are made.

[b] If defect is due to too-rapid freezing of material at the gate, mould temperature should be raised; if excessive shrinkage—due to too high temperature of melt in the mould—the mould temperature should be lowered.

a general guide, the ratio of re-grind to virgin material should preferably not exceed 1:4 (e.g. with glass-reinforced polyacetal, nylon 6.6, polyurethane) or even 1:5 (polycarbonate). If the re-grind is not of good quality, the maximum proportion added should be reduced still further. It is usual for the moulder to consult his material supplier in cases of doubt.

7.2.3.4 COMMON MOULDING FAULTS

The moulding faults which may occur in reinforced thermoplastic parts are listed in Table 7.6 together with the usual remedies. Most of these faults are encountered also in unreinforced mouldings, but some may be aggravated by the more rapid setting and higher viscosity of reinforced melts (e.g. short shots, undersize mouldings, weak weld lines, burning). Some faults may arise largely or entirely because of the presence and/or orientation of fibrous reinforcements (e.g. poor surface finish, warping of thin parts).

REFERENCES

1. Williams, J. C. L., Wood, D. W., Bodycot, I. F. and Epstein, B. N. (1968). 23rd Conference, RP/C Division, SPI, Paper 2-C.
2. Osborne, A. D. and Maine, F. W. (1974). 29th Conference RP/C Institute, SPI.
3. *SPI Handbook of Technology and Engineering of Reinforced Plastics/Composites* (1973). (Ed. J. G. Mohr), 2nd edn, Van Nostrand Reinhold, New York.
4. Lesseliers, L. (1972). 27th Conference RP/C Institute, SPI, Paper 14-E.
5. Sowa, M. W. (1970). 28th ANTEC SPE, Session 33, p. 703.
6. *Penn's PVC Technology* (1971). (Ed. W. V. Titow and B. J. Lanham), Applied Science Publishers Ltd., London, Wiley, New York.
7. Bernardo, A. C. (1970). 28th ANTEC SPE, Session 30, p. 629.
8. Mandy, F. (1974). Paper presented at the Semaine des Injecteurs, Brussels, Belgium.
9. Gluck, M. L. (1967). 22nd Conference RP/C Division SPI, Paper 8-A.
10. Norwalk, S. (4th October, 1971), RETEC SPE, Cleveland, Ohio.
11. Sachawa, J. C. and Slayton, J. L. (1972). 27th Conference RP/C Institute SPI, Paper 14-F.
12. Taylor, B. A. (1974). *Plastics and Polymers,* **42,** 157, 11.
13. Christopher, W. F. and Fox, D. W. (1962). *Polycarbonates,* Reinhold, New York, p. 83.
14. Schnell, H. (1964). *Chemistry and Physics of Polycarbonates,* Interscience, New York and London, p. 125.
15. Backofen, W., Vogel, H. and Kaminski, A. (1969). *Plastverarbeiter,* **20,** 8, p. 554.
16. *Zytel Glass-reinforced Nylon Resins* (1972). Moulding Manual, DuPont.
17. Taylor, R. B. (1970). *Plastics Technology,* **16,** 7, 48.
18. *Techniques for the Injection Moulding of Tefzel:* Technical Bulletin PIB No. 7, Du Pont de Nemours International S.A. (September 1972).

19. Gray, T. F., Jr. (1971). 26th Conference RP/C Division SPI, Paper 3-C.
20. Fiberfil Thermoplastic Moulding Compounds, Instruction Sheet No. R 3a, Ciba(ARL) Ltd., Duxford, Cambridge, England, July 1967.
21. Lucius, W. (1973). *Kunststoffe,* **63,** 6, 367.
22. Arpylene Reinforced Thermoplastic Moulding Materials: Processing Data Sheet FA 16, TBA, Reinforced Plastics Division, Rochdale, England, October 1972.
23. Sempert, R. (1973). *Kunststoffe,* **63,** 6, 365.
24. Pelka, H. and Vogel, H. (1968). *Plastverarbeiter,* **19,** 1, 10.
25. Pelka, H. and Vogel, H. (1967). *Plastverarbeiter,* **18,** 12, 837.
26. Filbert, W. C. (1969). *SPE Journal,* **25,** 1, 65.
27. Maxwell, J. (1964). *Plastics Today,* No. 22, October, 9.
28. Fucella, D. C. (1972). 27th Conference, RP/C Institute SPI, Paper 11-F.
29. Schmidt, L. R. (1974). 'Injection Moulding of Glass-Filled Polypropylene', Paper presented to the meeting of the Polymer Physics Group of the Institute of Physics, July 9th–12th, Nottingham, England.
30. Speeuwers, H. R. (1970). Plastics Institute Conference on Reinforced Thermoplastics, October, Solihull, England, Paper 7.
31. Olmsted, B. A. (1970). *SPE Journal,* **26,** 2, 42.

CHAPTER 8

APPLICATIONS AND DEVELOPMENTS

8.1 CURRENT FIELDS OF APPLICATION AND USES OF REINFORCED THERMOPLASTICS

A general indication has already been given in Chapter 1, Section 1.2. of the major fields of application of commercial reinforced thermoplastic materials and of the approximate proportional consumption in these fields of glass-reinforced thermoplastics. In terms of tonnage the latter materials are the most important single group of reinforced thermoplastics (about 90%), with glass-fibre-reinforced nylon and polypropylene still the biggest sub-groups in, respectively, Europe and the USA.

The main established applications are mentioned in more detail in this section. No attempt has been made to provide a complete list, or to group the applications under the headings of individual polymers or polymer-reinforcement combinations. Neither would have been practicable, the former because of the number and diversity of applications, and the latter because different reinforced thermoplastics are often used, and compete, in the same application. However, the types of application mentioned, and the examples given to illustrate them, should provide a reasonable outline of the current usage of reinforced thermoplastics. In many cases the particular materials associated with an application are indicated, in brackets, in abbreviated form. The abbreviations are the same as used (and explained) in Figs. 6.2–6.5 (see Appendix 2); they are in any case largely self-explanatory. Where a material is shown in conjunction with a group of products the implication is that it may be used in some, but not necessarily all, of the products in the group.

A few general points may also be mentioned, concerning some of the important property features in the applicational context. Thus flame retardant grades of reinforced thermoplastics are used particularly widely in the electrical applications, to reduce fire hazards. Similarly, low-friction compounds find a major outlet in bearings, gears, cams and the like in the business machine, appliance and engineering field. These two special types of compound are discussed in Chapter 9. The surface finish of many reinforced (in particular glass-fibre-reinforced) thermoplastic parts, whilst it may be reasonably good by ordinary standards (especially when high injection speed and a hot mould are used in moulding—*cf.* Chapter 7), is not generally suitable for electroplating. Fybex-reinforced mouldings, especially in ABS, PPO and polypropylene, attain the necessary standard of surface finish; this is an important consideration in the use of this reinforcement. Plating is an important finish in certain car parts (e.g. grilles) record player parts (e.g. tone arms), and plumbing components. Whilst the precision of moulding such features as screw threads in a reinforced thermoplastic is comparable with that in, say, die-casting of aluminium, superior resistance to stress rupture and/or fatigue may be obtained.

8.1.1 Automotive Applications

These were developed earlier—and are still more numerous and greater in terms of material consumption—in the USA, but development and expansion in Europe and the UK are now vigorous. For example, the increase in total annual consumption of glass-reinforced thermoplastics for exterior painted parts from 1973 to 1974 is estimated by the Owens-Corning Fiberglas Corporation as about seven-fold (from 700 000 to 5 million lb[1]).

The main areas of application with a few typical examples are given in Table 8.1 (pp. 243–5). Some applications are illustrated in Figs. 8.1–8.6. Owens-Corning Fiberglas lists some 150 separate applications of glass-fibre-reinforced thermoplastics in motor cars for 1974.[1]

In American cars in particular, reinforced styrenics are used in large quantities for interior components, e.g. dashboards, consoles, panels and crash-pad inserts. The scale of this type of application is so far very much smaller in Europe. Another fast-growing (and now extensive) mainly American application is that of glass-reinforced polypropylene for lamp housings. Polypropylene reinforced with asbestos has been used on a substantial scale in the UK for heater boxes and duct systems, and also—in both the UK and Europe—in radiator fan shrouds, rear parcel shelves, ventilator and air-inlet valves. Some parts are being made also in talc-reinforced polypropylene.[2–4]

8.1.2 Tools: Appliances: Engineering

Power tools are an important application in this group: in accordance with established usage patterns power tool bodies and handles are moulded mainly in nylon 6 in Europe and nylon 6.6 in the UK. Other noteworthy applications include the use of glass-reinforced polypropylene in applications where resistance to hot water is important, e.g. pump components, impellers in washing machines, and the use of glass-

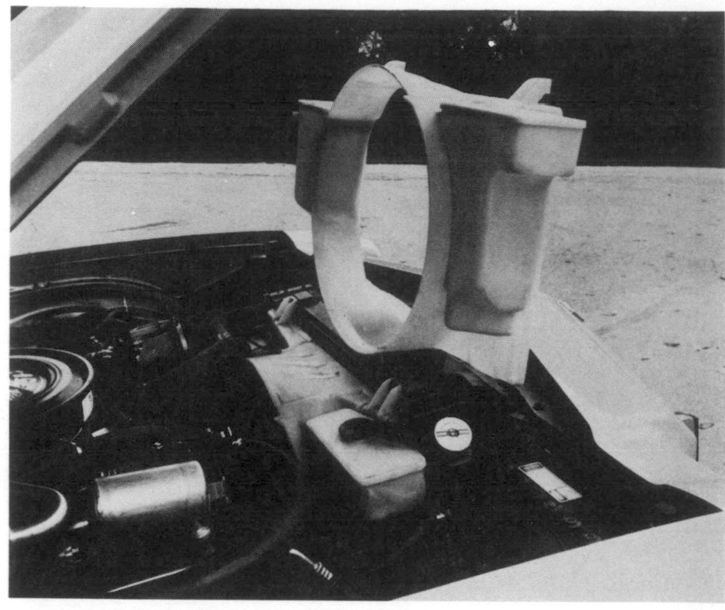

FIG. 8.1 Injection-moulded polypropylene/glass fibre (20 wt %) component: combined fan shroud, storage tank for windshield washers and radiator overflow tank. Used on all Buick models for 1973. (Courtesy of Mr. F. Mandy and Owens-Corning Fiberglas Europe S.A.)

reinforced low-friction materials (often internally lubricated), *viz.* nylon, acetal polymers, thermoplastic polyesters, for cams, gears and levers.

The main types of application are listed below, with some specific examples:

Power tools: housings and handles (N 6.6/GF or 6/GF; PC/GF; PBT/GF); bushes and sleeves (N 6.6/GF or 6/GF).

Hand tool handles: hammer and screwdriver handles (N 6.6/GF); exchangeable-blade and electric knife handles (N 6.6/GF; PBT/GF); paint brush handles (PP/GF structural foam); secateur handles (N 6.6/GF).

Domestic appliance parts: washing machine lids, casings and parts (PPO/GF; PP/T; N 6.6/GF or 6/GF); washing machine side-check valves (SAN/GF); pump and impeller housings (PP/A); pump bodies and/or impellers (PPO/GF; PP/GF); dish-washer sprinklers (PPO/GF); floor polisher head parts (PS/GF); hair dryer components (PP/A).

Water engineering and plumbing: submersible pumps or components (PPO/GF); tap and valve parts (PPO/GF); water-meter housings (PC/GF; PPO/GF; PBT/GF; N 6.6/GF); filter units or components

(ABS/GF; PA/GF); water-softener or mixing valves (PBT/GF; PPO/GF); valve controls (SAN/GF).

Working parts: gears, bearings and bearing cages, cams, levers (PBT/GF; N 6.6/GF or 6/GF or B or C or F; PA/GF).

Miscellaneous: air-conditioning propeller fans (PPO/GF) and housings (PPO/GF; PP/A); conveyor rollers (HDPE/B); parking meter spring

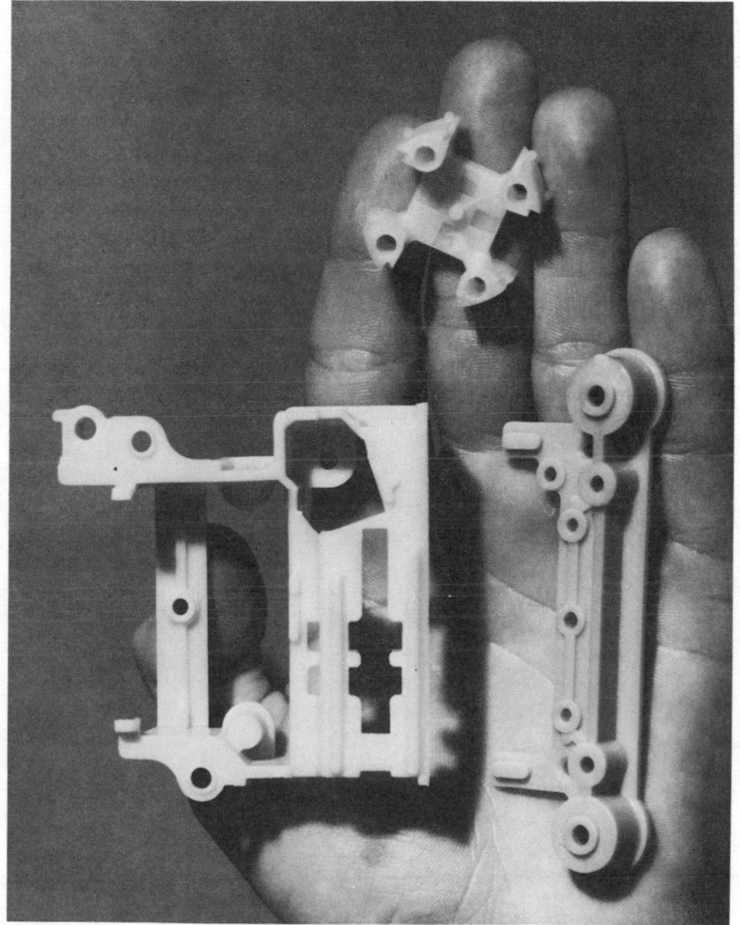

FIG. 8.2 Working components for General Motors 1972 car models. Oil-pressure gauge bobbins (circular part in the foreground): injection-moulded glass-fibre reinforced nylon 6.6 Jewel mounts (centre and top) for instrumentation systems: acetal polymer/glass fibre (30 wt %). (Courtesy of Mr. F. Mandy and Owens-Corning Fiberglas Europe S.A.)

FIG. 8.3 Rear end finishing panel (Pontiac Le Mans). Thermoplastic polyester reinforced with 30 wt % glass fibre. (Courtesy of Mr. F. Mandy and Owens-Corning Fiberglas Europe S.A.)

FIG. 8.4 Intake air cleaner (Ford Pinto). Polypropylene reinforced with 20 wt % glass fibre. (Courtesy of Mr. F. Mandy and Owens-Corning Fiberglas Europe S.A.)

(a)

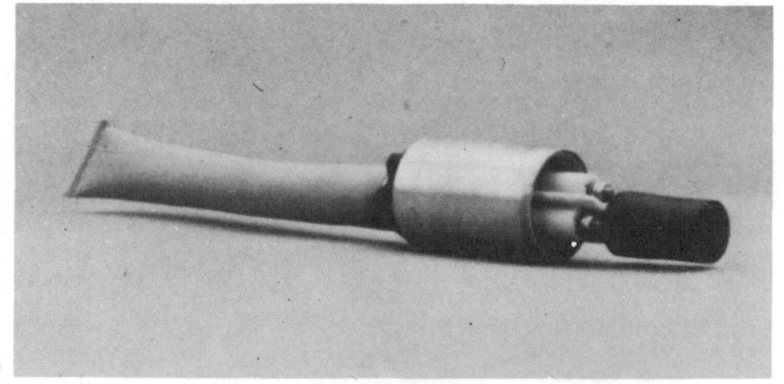

(b)

FIG. 8.5 Components of the 1972 Chevrolet Vega. (a) Fuel pump injection moulded in acetal polymer (65%) with PTFE (5%), reinforced with glass fibre (30%). (b) Timing belt cover polypropylene/glass fibre (20 wt %). (Courtesy of Mr. F. Mandy and Owens-Corning Fiberglas Europe S.A.)

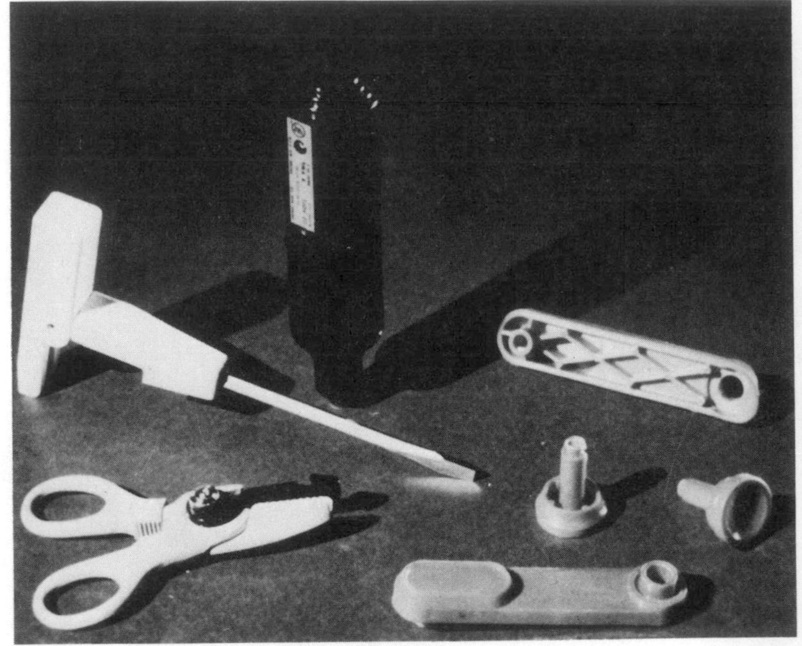

FIG. 8.6 Flower shears, combination screw driver/hammer, power tool housing, and components, moulded in nylon with 33% glass fibre reinforcement. (Courtesy of Mr. F. Mandy and Owens-Corning Fiberglas Europe S.A.)

housing and money box (PPO/GF); window sashes and stays (N 6.6/GF or 6/GF); tie brackets and bars for concrete forming (N 6.6/GF).

8.1.3 Electrical and Electronic (including computers)

This is a large outlet, particularly for glass-reinforced nylon and polycarbonate. The applications which are numerous and varied include some very small components, e.g. coil-formers, bobbins, connectors. In these and other applications many components traditionally produced in thermosetting materials are now made in reinforced thermoplastics. The following are the common types of application, with examples:

Housings and spools: computer and electronic equipment housing components, meter housings (PC/GF; PPO/GF); tape spools (SAN/GF).

Boxes, covers and panels: terminal, fuse and junction boxes; guard panels (PC/GF; N 6.6/GF; PBT/GF); battery cases (HDPE/B).

Various small components: coil formers and bobbins, fuse blocks and sockets, plugs and connectors, terminal blocks, switches, switch housings

TABLE 8.1

Applications of Reinforced Thermoplastics in Motor Cars

(Based in part on information published by the Owens-Corning Fiberglas Corporation)

RTP material			Examples		
	Reinforcement				
Base polymer	Nature	Wt %	Component	Approximate weight (lb)	Car make
1. *Functional Components*					
Polypropylene	Glass fibre	30	Fender apron	9	Ford: Thunderbird Continental MK IV
Polypropylene	Glass fibre	30	Hose elbow	0·33	Fiat: 126; 500
Polypropylene	Glass fibre	20	Fan shroud	3·9	Buick: Century: Regal
Polypropylene	Glass fibre	30	Heater housing assembly	13·2	Opel: Kadett
Polypropylene	Glass fibre	15	Torque converter cover	0·8	Oldsmobile: Cutlass: Delta 88
Polypropylene	Talc	20	Heater casing	—	Vauxhall: Viva
Polypropylene	Asbestos	40	Heater casing	—	British Leyland: Marina
Nylon	Glass fibre	50	Heater housing levers	0·29	Fiat: 126; 500
Nylon	Glass fibre with glass spheres	—	Fuse box	0·4	BMW 520
Nylon 6	Glass fibre	60	Air-intake duct	2·45	Volkswagen: 411
Nylon 6	Glass fibre	30	Inlet pipe	1·76	Porsche
Nylon 6.6	Glass fibre	33	Valve stem oil seals	0·01	Ford: all cars; GM: all cars
Nylon 6.6	Glass fibre	30	Door lock system	0·44	Citroen GS
Acetal	Glass fibre	20	Mini-vent window frame	1·1	Lincoln Town Car
Acetal	Glass fibre	20	Speedometer jewels	0·03	GM: all cars
Acetal (65%) PTFE (5%)	Glass fibre	30	Electric fuel pump	0·2	Chevrolet: Vega
PPO (Noryl)	Glass fibre	20	Pipe fittings	0·011	Renault: all cars
PPO (Noryl)	Glass fibre	20	Heat tap plug	0·033	Renault: 12
Polystyrene	Glass fibre	20	Air-conditioner adapter	0·6	Cadillac

Table 8.1—*continued*

RTP Material			Examples		
	Reinforcement				
Base polymer	Nature	Wt %	Component	Approximate weight (lb)	Car make
2. *Consoles and Interior Trim Accessories*					
Nylon	Glass fibre	30	Speedometer frame	0·44	Citroen 6S
SAN	Glass fibre	20	Consoles (lower and upper)	4·25	Oldsmobile: Cutlass
SAN	Glass fibre	20	Door console	0·3	Oldsmobile: Cutlass
SAN	Glass fibre	20	Tape player case assembly	0·7	Oldsmobile: Cutlass
ABS	Glass fibre	20	Seat separator console	5·0	Chevrolet: Chevelle, Monte Carlo
ABS	Glass fibre	13	Glove box door	0·3	Chevrolet: Camaro
Polystyrene	Glass fibre	20	Glove box door base	1·1	Cadillac and Eldorado
Polystyrene	Glass fibre	20	Ash tray door	0·9	Cadillac and Eldorado
3. *Lamp Housings and Accessories*					
Nylon 6.6	Glass fibre	30	Front lamp housing	2·2	Citroen DS
Nylon 6.6	Glass fibre	30	Tail lamp housing	0·66	Peugeot: various models
Polypropylene	Glass fibre	20	Tail lamp housing	0·5–1·79	Various American makes and models
Polycarbonate	Glass fibre	8	Head lamp bezel	1·6	Oldsmobile: Cutlass
ABS	Glass fibre	20	Tail lamp housing	1·0	Dodge: Charger
ABS	Glass fibre	20	Side marker assembly	0·13	Chrysler: all models

Table 8.1—*continued*

RTP material			Examples		
	Reinforcement				
Base polymer	Nature	Wt %	Component	Approximate weight (lb)	Car make
4. *Instrument Panels and Accessories*					
Polypropylene	Talc	20	Dashboard	1·5	Fiat: Autobianchi A112
Polypropylene	Glass fibre	20	Panel assembly upper quarter	3·1	Buick: Riviera
ABS	Glass fibre	13	Instrument panel	2·8	Chevrolet: Camaro
ABS	Glass fibre	13	Instrument panel cluster bezel	5·0	Am. Motors: Matador, Ambassador
ABS	Glass fibre	20	Instrument panel and retainer	8·5	Chevrolet: Chevelle, Monte Carlo
Polystyrene	Glass fibre	20	Instrument panels and panel retainers	3·7–5·7	Some Oldsmobile and Pontiac models
SAN	Glass fibre	20	Instrument panels	up to 7·5	Pontiac, Oldsmobile, Chevrolet, Cadillac
5. *External Components*					
Polyester	Glass fibre	30	Rear-end panels	up to 3·5	Pontiac, Oldsmobile, Chevrolet
Nylon	Glass fibre	15	Rear-end panel	4·2	Oldsmobile: Omega
Nylon	Glass fibre	30	Front grille	0·88	Mercedes-Benz
Polypropylene	Glass fibre	30	Louvre	0·44	Opel: Manta
Polyester	Glass fibre	30	Rear side window louvre	4·7	Pontiac: Grand Am.

and circuit breakers (PC/GF; N 6.6/B; N 6.6/GF or 6/GF; PPO/GF; PPO/B; PBT/GF; PP/A; N 6.6/A); support insulators (PBT/GF); cable duct connectors (N 6/GF); electric shaver parts (N 6.6/GF).

Television and radio parts: tuners and tuner parts (PBT/GF; PPO/GF); T.V.-set back panels (PP/A); car radio brackets (PC/GF).

Light fittings: street lamp housings (N 6.6/GF); other light housings and fittings (PP/GF or A; PC/GF).

Electric motor parts: motor fan rotors and blades (N 6/GF or 6.6/GF; PP/A); motor end-caps and commutators (PBT/GF; N 6.6/GF); switch gear housings (PP/GF); motor support panels (PPO/GF).

8.1.4 Business and Office Machines

These provide a considerable outlet for reinforced thermoplastics.

Housings and casings: cowls, covers, panels and casings for calculators (PC/GF; PPO/GF) and copying machines (SAN/GF).

Internal mechanism: as 'working parts' under Section 8.1.2 above.

8.1.5 Miscellaneous

Chemical plant: pumps, pump housings and parts (PP/GF; PA/GF; PBT/GF); pipe flanges (PP/GF or A); filter units (PP/GF); packing rings, valves (PP/GF; ETFE/GF); gas valves (PBT/GF; N 6.6/GF).

Photographic and cine equipment: camera housings (PC/GF; N 6.6/B); projector housings (PC/GF; ABS/GF; PPO/GF).

Furniture and fittings: chairs (PP/GF—solid or structural foam); cabinet doors (PS/GF structural foam); drawers (ABS/GF); shelf brackets and standards (PS/GF; SAN/GF; N 6.6/GF*).

Small components and articles: watch cases and back plates (N 6.6/GF; PPO/GF; N 12/B); watch-works mounting plates (N 6.6/GF); gauge plates (PSu/GF); poultry watering cups (ABS/B); brush blocks (PP/GF structural foam); pharmaceutical beakers (PBT/GF).

Various products: welding hood components (PP/A); fruit crates (HDPE/GF structural foam); baseball bats (N 6.6/GF); edge-trim profiles and panels (PVC/F); missile nose cones (PPO/GF); sailing dinghy (ABS/GF); portable record player casing (PS/GF) and speaker box (PP/GF); pipe coupling sleeves (SAN/GF); heat insulating liners (N 6.6/A); motor cycle mudguards (N 6/GF); textile bobbins, yarn guides, pulleys (PS/GF; ABS/GF; N 6.6/GF or 6/GF).

* Material used for hat-rack support bracket on the Concorde supersonic aircraft.[5]

8.2 PERFORMANCE SPECIFICATIONS

Many companies and some official bodies draw up specifications for materials, products and components; some of these cover, or extend to, reinforced thermoplastics. Whilst these specifications are primarily of interest to suppliers, they do set up performance standards whose existence is relevant in the general context of the applications and uses of reinforced thermoplastic materials.

The specifications fall into the following general groups:

(a) military (ordnance) specifications in various countries;
(b) government and official specifications; and
(c) individual company specifications.

Among the military specifications, several US Specifications relate to reinforced thermoplastics (mainly glass-fibre-reinforced), as do the British Specifications, issued by the Directorate of Quality Assurance (Materials).

Other official specifications are exemplified by the US Federal Specifications including the Federal Motor Vehicle Safety Standards some of which cover motor car parts and components. Some of the regulations of the Economic Commission for Europe also apply in this field.

Motor car materials and parts form the subject of a large number of the 'individual user' specifications, established by all the major motor vehicle manufacturers. Such individual company specifications are also used by manufacturers in the major areas mentioned under Section 8.1 above, especially many computer companies and those producing business machines and domestic appliances.

8.3 SOME RECENT AND CURRENT DEVELOPMENTS

Comparatively recent developments in commercial reinforced thermoplastic materials include reinforcements newly available or expected to become available as well as polymers which have only recently appeared as base resins in commercial reinforced thermoplastics.

8.3.1 New Reinforcements for Commercial Reinforced Thermoplastics

The appearance of carbon fibres in commercial reinforced thermoplastic compounds is so recent that it could be considered under this heading. However, because the fibres themselves have been available for a considerable time, they are discussed in Chapter 4. Two current topics in this area are the present work on improvements in carbon fibres produced from pitch (Union Carbide) and the continuing development of coupling treatments for carbon fibres used in reinforced thermoplastics.[6]

The next new fibrous reinforcement expected to find its way into thermoplastics will probably be one or more of the high-modulus organic fibres, so far used only in thermoset composites. The generic name 'Aramid', now approved by the US Federal Trade Commission, applies to the aromatic polyamide fibres of this general group, whose most commercially advanced members are based on poly(*p*-phenylene tetraphthalamide) and poly(*p*-benzamide) developed by DuPont, and on aromatic polyamide and polyamide-hydrazide polymers developed by Monsanto.[7–9] Their respective designations are Kevlar and X-500.

The modulus, strength and density values typical of such fibres are given in Table 8.2, in comparison with those of steel wire and drawn silica fibres (the strongest fibres in the glass family). The fibres retain their properties in short-term exposures up to about 300°C; some do not decompose up to 500°C. Depending on the actual polymer, and the processing in the course of production, they can have extensibilities several times higher than those of carbon fibres.

With the recent drop in the price of Kevlar 49 (formerly PRD-49) this fibre may well become the first of its type to be used as reinforcement in thermoplastics.

Special grades of surface-treated glass fibre, for use as reinforcement in thermoplastics have appeared, some specifically for particular base polymers: e.g. glass fibre Series 411 and 409 (Owens-Corning) or CS 704 (Johns Manville) for nylon and polycarbonate.

A few non-fibrous fillers—said to improve some properties—have recently been finding their way into commercial thermoplastic compounds. Whatever the real extent of property improvements in particular cases, there is no doubt that, being cheaper than the polymer they replace in a compound, these fillers are economically useful extenders; the clays (see below) have certainly served this purpose in rubbers in which they were first used many years ago.

As they are particulate, their effect on mechanical properties is isotropic.

TABLE 8.2

Stiffness and Strength of the High-modulus Organic Fibres (average or typical values) in Comparison with other High Strength and Modulus Filamentary Materials

Material	Specific gravity	Young's modulus (psi × 10^6)	Tensile strength (psi × 10^3)
High modulus organic fibres	1·44–1·47	12–19	200–400
Drawn silica	2·5	10–12	750–850 (ultimate)
Steel wire	7·8	30	300–400

TABLE 8.3
Clay-filled Nylon 6.6 (dry as moulded, 40% by weight filler): Reinforcement and/or Change Factors for some Properties (base polymer = 1 in each case)

Property	Flexural modulus	Tensile strength	Flexural strength	Deflection temp. at 264 psi	Coefficient of linear thermal expansion	Izod impact strength
Factor	2·5	1·2	1·5	3·1	0·66	0·9

The fillers are wood flour (processed to ensure stability in hot plastics melts) used in rigid PVC and polypropylene sheeting and profiles,[10] and clays used in nylon moulding compounds. The wood-filled sheets find application as automotive interior panels, in furniture and in building interiors: the last-named application is also the main one for the profiles. The clay-filled nylons are claimed to be suitable for many applications in which reinforced nylons are used.[11]

The effects of the clay fillers on some key properties of nylons are given—in terms of reinforcement or change factors—in Table 8.3.

8.3.2 Recent Commercial Base Polymers

Two thermoplastic polymers may be mentioned under this heading, both now available commercially in reinforced form. These are polyphenylene sulphide, and polyimides. The glass-reinforced versions of these polymers, e.g. the LNP Thermocomp OF and YF series respectively, are recommended for applications requiring high-temperature stability and chemical resistance.

A thermoset polyester supplied (with glass reinforcement) in pellet form, and suitable for injection-moulding, has been available since 1972 from Fiberfil.[12] This material thus processes like a thermoplastic (remaining stable in the barrel of an injection machine at about 125°C) but cross-links rapidly at about 165°C (in a heated mould), to give parts with heat distortion temperatures up to about 200°C.

8.3.3 Some Recent Product Developments

Only a few items will be mentioned by way of example.

Gears for petrol pumps have recently been moulded in glass-fibre-reinforced polyphenylene sulphide, whilst the applications of Fybex-reinforced nylon in working parts were increasing up to 1974.

Golf-club faces and tennis racquets have been produced in nylon reinforced with carbon fibre.

Continued growth is expected in the applications for reinforced structural foam components, and reinforced thermoplastic polyesters for components with good creep properties at elevated temperature.[13]

Reference has already been made (in Chapter 7) to the recent development of composite sheet materials based on thermoplastic polymers. The sheets, e.g. the STX™ composite introduced in 1974 by the Allied Chemical Corp., are intended for fabrication by stamping, hot or cold.

REFERENCES

1. 'Fiberglas/Plastics in the 1974 Automobile', Publication No. 5-AU-6340, Owens-Corning Fiberglas Corporation, November 1973.
2. Harrison, P. and Sheppard, R. F. (1970). 'Reinforced Polypropylene', Paper presented at The Plastics Institute Conference on Reinforced Thermoplastics, Solihull, England, October.
3. Hill, A. C. (1973). 'Reinforced Thermoplastics in Automobile Applications', Paper presented at the conference on Advances in Reinforced Thermoplastics, Kingston Polytechnic, Kingston, Surrey, England, April.
4. Anon. (1973). *Modern Plastics International,* November, 20.
5. Anon. (1971). *Modern Plastics International,* September, 17.
6. Anon. (1973). *Modern Plastics International,* December, 60.
7. DuPont US Patents 3 536 651 (1970); 3 595 951 (1971); 3 600 350 (1971); 3 642 707 (1972).
8. Black, W. B. and Preston, J. (Eds.) (1973). *High Modulus Wholly Aromatic Fibres,* Marcel Dekker, New York.
9. Chiao, T. T. and Moore, R. L. (1973). *Composites,* **4,** 1, 31.
10. Anon. (1974). *Modern Plastics International,* February, 22.
11. *Ibid.,* August 1972, 32; November 1972, 40.
12. *Ibid.,* June 1972, 32 and 46.
13. Best, J. R. and Kelly, T. C. (1974). 29th Conference, RP/C Institute SPI, Paper 24-G.

CHAPTER 9

SPECIAL REINFORCED THERMOPLASTIC MATERIALS

9.1 GENERAL

There are two kinds of reinforced thermoplastic materials of commercial importance which may be described as 'special' by virtue of their composition and application. These are flame retardant and low-friction thermoplastic materials. Each of the two groups contains some compounds which possess the special characteristics, at least in some degree, by virtue of the 'natural' inherent properties of the base polymer or polymer/reinforcement combination: this applies particularly to the flame retardant thermoplastics. Such compounds are, however, in the minority and in both groups the extra properties required are predominantly imparted by special additives.

9.2 FLAME RETARDANT REINFORCED THERMOPLASTICS

The considerations of safety in use, and in particular the need for low flammability of reinforced thermoplastics, have grown in importance with the increasing application of these materials in the building industry, in electrical components, portable tools, domestic and office appliances.

The concern of the users (fabricators) of reinforced thermoplastic materials, and the consumers of the finished products, for the greatest possible safety in regard to fire hazards has not only been reflected in the emergence of evaluation tests, but also has resulted in legislation (strongest in the United States, some in other countries) aimed at ensuring reasonable performance standards in this respect.

This attitude and the requirements made it imperative that tests should be available to predict the flame resistance in service of reinforced thermoplastics.

9.2.1 Flammability Test Methods and Rating

The commonly used standard test methods include those of the following specifications:

ASTM D-635
The Underwriters Laboratories Subject 94 (Part B)
IBM CMH 6-0430-102
ASTM D-2863: The 'Oxygen Index' method based on the Fennimore–Martin flammability test[1]
BS 476 Part 7
BS 4735

The first four tests are summarised in Table 9.1. Another summary, with special reference to the oxygen index test was recently published by Abbott.[2]

The two British Standard methods use horizontal samples. In general, tests carried out with the sample in this position are less severe than those in which the sample is vertical.

Another test is also used in Europe to test the flammability of commercial materials and products, in particular for electrical applications (including electrical household appliances). This is the 'hot wire' or 'glow wire' test[3-5] which is the subject, *inter alia*, of the following British and German Specifications:

BS 3456: Section 2.10: 1972
VDE* 0471 Part 2—70

In the test the tip of an electrically heated wire of special configuration,[3] maintained at 960°C, is applied to a point near the centre of the surface of the material or article to be tested. In the course of the determination the wire is horizontal and the surface vertical. The specimen is pressed against the wire tip by the force of 1 N for a period of 30 s. The specimen fails the test if:

(a) the length of any flame arising whilst the wire tip is in contact with the specimen exceeds 30 mm; or
(b) the flame persists for more than 30 s after the wire tip has been removed from the specimen; or
(c) the depth of penetration of the wire tip into the specimen is greater than 15 mm.

Special tests to assess susceptibility to ignition by 'electrical' sources have also been developed by Underwriters Laboratories Inc.[6]

The UL Subject 94 test is the one most important commercially: most of the users of plastics materials, including reinforced thermoplastics (and in particular appliance manufacturers), will not accept materials for the production of their manufactured products unless the material has a UL test rating. The ratings shown in Table 9.1 as 'vertical group' ratings (Classes 0, I and II) were originally known as 'self-extinguishing' ratings, and references to, say, 'SE-I' or 'SE-0' are frequently encountered in recent and even current literature and specifications. However, the correct

* German Electrical Engineers Association.

TABLE 9.1
Common Flammability Tests
(Courtesy of Liquid Nitrogen Processing Corporation)

	ASTM D-635	UL Subject 94 (Part B)[a]	IBM CMH 6-0430-102[c]	Oxygen index method [b]ASTM D-2863T
Test specimen	Bar 5 in × $\frac{1}{2}$ in × $\frac{1}{8}$ in (or $\frac{1}{4}$ in)	Bars 6 in × $\frac{1}{2}$ in × $\frac{1}{16}$ in and 6 in × $\frac{1}{2}$ in × $\frac{1}{4}$ in	Bar 5 in × $\frac{1}{2}$ in × $\frac{1}{8}$ in	Bar 7 to 15 cm long × 6·5 ± 0·5 mm wide × 3·0 ± 0·5 mm thick
Number of specimens	10	3	5	10
Flame characteristics	1 in Height, blue flame	$\frac{3}{4}$ in Height, blue flame	0·75 in Yellow-blue flame	6–12 mm long
Sample position	Horizontal	Vertical	Vertical	Vertical
Position of flame to sample	Tip of flame contacts end of specimen	Specimen $\frac{3}{8}$ in above burner tube	Tip of flame contacts end of specimen. Burner may be tipped 20° from vertical	Ignite top of sample
Ignition time	30 s	10 s	5, 10, 15, 20, 25, 30, 35, 40, 45, 50, 55 and 60 s	Until entire top is burning
Number of ignitions per sample	2	2	1	1
Procedure	Apply Bunsen burner flame to tip of specimen for 30 s. If it does not ignite, wait 30 s and apply again for 30 s	Apply Bunsen burner flame to lower tip of specimen for 10 s. If it does not ignite or stops burning before the end of a 30 s waiting period, apply again for 10 s immediately after sample has stopped glowing	Apply Bunsen burner flame to tip of specimen for 5 s. If specimen does not ignite or stops burning before the end of 30 s waiting period, apply Bunsen burner flame to additional specimen for 10 s, increasing flame application time in 5 s increments on fresh specimens up to 60 s	Ignite top of specimen so that entire top is burning. Remove flame and start timer. (i) The concentration of oxygen is too high and must be reduced if: (a) specimen burns 3 min or longer, or (b) specimen burns 5 cm. (ii) The concentration of oxygen must be raised if the specimen is extinguished before burning 3 min or 5 cm. Continue to adjust oxygen content until the limiting concentration is determined

Table 9.1—*continued*

	ASTM D-635	UL Subject 94 (Part B)[a]	IBM CMH 6-0430-102[c]	Oxygen index method [b]ASTM D-2863T
Classification	Burning—the sample burns more than 1 in; the burning rate is reported in in/min (note—it is common practice to omit the actual burning rate and report the result as 'slow burning') Self-extinguishing—the sample ignites but burns less than 1 in Non-burning—the sample does not ignite	Vertical, group: 0 does not burn for more than 10 s (an average of 5 s for 5 samples), does not burn up to the clamp, does not drip flaming particles which ignite surgical cotton, and does not glow beyond 30 s after the second flame application (10 s after first application) I does not burn for more than 30 s (an average of 25 s for 5 samples), does not burn up to clamp, does not drip and ignite cotton, and does not glow beyond 60 s, after second flame application II same as (I) above except sample drips and ignites the cotton and when tested in accordance with slow burning procedure it does not burn 4 in	Class A—Materials which after a 60 s flame application time will extinguish within 30 s without producing flaming droplets Class B—Materials which after a 5 s or more flame application time will extinguish within 30 s without producing flaming droplets, but cannot withstand a 60 s flame application Class O—Materials which burn for more than 30 s or produce flaming droplets that ignite cotton after a 5 s application	The limiting oxygen concentration is defined as that which will meet conditions (i)(a) above. At the next lower concentration the specimen should extinguish as defined in (ii). Calculate oxygen index (n) as follows: $n(\%) = \frac{[O_2] \times 100}{[O_2]+[N_2]}$ The higher the oxygen index of a material, the greater the flammability resistance

[a] Part A similar to ASTM D-635; Part B applied only to materials not rated as 'burning' by Part A.
[b] Fennimore–Martin flammability test.
[c] Extensively revised in 1974: now closely in line with the UL Standard.

current designation is the one shown in Table 9.1 and the ratings are abbreviated to V-0, V-I and V-II. This recent change in nomenclature has been brought about by the activities of various consumer groups because of the possible misunderstandings over the term 'self-extinguishing'. People not closely acquainted with the UL specification and test might tend to interpret this designation literally, i.e. as meaning that a material within this classification would not burn at all. In the new designation the adjective 'vertical' refers to the position of the sample (*cf.* Table 9.1) which makes the test more severe than tests in which the sample is placed horizontally.

Almost all industrial users insist on a suitable UL rating (in many cases V-0 or V-I) of the reinforced thermoplastic materials they process, because their products in turn have to meet similar requirements.

The world-wide interest among processors and users in the flammability resistance of plastics materials stemmed in large measure from the activities of the Underwriters Laboratories. The interest is, if anything, even keener in the reinforced thermoplastics sector of the industry and, particularly in the USA, most appliance manufacturers will normally refuse to employ materials which have not been given UL approval in the form of an acceptable rating. The UL test is currently regarded as the most reliable and probably the most severe of the common tests, although the Oxygen Index test adopted by the ASTM affords the basis for a more quantitative rating. Additionally, individual companies and organisations have developed their own tests or adapted existing tests to their requirements.

The general way in which the UL scheme is applied in materials selection and acceptance is as follows. The supplier of a particular grade of raw material or reinforced compound carries out the UL Subject 94 Part B test on his material. He then submits samples of the material, together with his test results, to the Underwriters Laboratories* who carry out their own determinations and allot a rating to the particular grade. It is to the advantage of a manufacturer using a number of materials in his product to be able to show that all the materials have UL recognition (yellow card) because the products will then be accepted more easily. It is usual to describe materials with UL Subject 94 ratings of V-0 or V-I as flame retardant.

9.2.2 Commercial Flame Retardant Reinforced Thermoplastics

9.2.2.1 Flammability Characteristics of Commercial Materials

The incorporation in thermoplastics of glass fibres or asbestos fibres,

* Headquarters: Chicago, Ill., USA; several testing stations in the USA; representatives in the USA and abroad (including the UK—British Standards Institution, London).

TABLE 9.2
Flammability Ratings of Some Commercial (Thermocomp) Reinforced Thermoplastics
(Courtesy of Liquid Nitrogen Processing Corporation)

Base polymer	Glass content (%)	LNP series code	Flammability test and rating (*cf.* Table 9.1) ASTM	UL	IBM[a]	Oxygen index[a]
	Units		—	—	—	% O_2
	ASTM		D635	Subj 94	CMH 6-0430-102	D2863T
SAN	30	BF-1006 FR	NB	V-0	A	26·5
Polycarbonate	30	DF-1006	SE	V-I	A	29·8
Polycarbonate	10	DFA-113	SE	V-0	A	31·5
Polycarbonate	30	DFL-4036[b]	SE	V-I	A	30·5
Polysulphone	30	GF-1006	SE	V-0	A	35·0
Polysulphone	30	GF-1006 FR[c]	SE	V-0	A	39·5
PPO	30	NF-1006	SE	V-0	A	28·0
Polypropylene	20	MF-1004 FR	NB	V-0	A	27·2
PVC (copolymer)	15	VF-1003	NB	V-0	A	40·0
Nylon 6	30	PF-1006 FR	NB	V-0	A	28·0
Nylon 6.10	30	QF-1006 FR	NB	V-0	A	28·0
Nylon 6.6	30	RF-1006 FR	NB	V-0	A	27·5
Polyester (PTMT)	30	WF-1006 FR	NB	V-0	A	27·0
Modified PPO	20	ZF-1004 FR	SE	V-I	B	27·2

FR = Special flame-retardant formulation.
SE = Self-extinguishing.
NB = Non-burning.
[a] ASTM Task Forces currently preparing test methods.
[b] DFL-4036—30% glass reinforced, 15% TFE lubricated polycarbonate.
[c] Low smoke generation grade.

which are the most common and widely used reinforcements in commercial materials, improves the burning resistance of the composite in comparison with the base polymer alone. The fibrous reinforcement conducts heat away from the site of combusion and minimises after-glow to which many unreinforced polymers are prone (even some partly flame resistant varieties). This effect is especially advantageous in a horizontal flame test such as the ASTM D-635 (where the entire test bar is not surrounded by the hot gases from the burner) and in situations of analogous exposure to flame in actual service. The fibre-to-resin bond also adds integrity to the composite, minimising dripping and decreasing the burning rate.[7] These flame retardant effects of the reinforcement can in some cases reduce the amount of the special flame retardant additives (see below) which may have to be added. Since the incorporation of such additives may affect processing and performance, the reinforcement materials are also indirectly beneficial in this way.

Some of the base polymers in reinforced thermoplastics are inherently flame retardant. When glass fibre reinforcement (or asbestos fibre if appropriate) is incorporated the flammability resistance can be improved still further in accordance with the points mentioned above. This is illustrated by the data of Table 9.2 for commercial grades of some reinforced thermoplastics (Thermocomp series: Liquid Nitrogen Processing Corp.). In the table only the grades with the suffix 'FR' are special flame retardant formulations. It can thus be seen that glass-reinforced PVC, polysulphone and PPO can attain the highest UL flame retardancy rating without special additives. Polycarbonate is only slightly inferior.

Table 9.2 also illustrates the comparison of ratings in various tests. As can be seen, the glass-reinforced PVC is the least flammable material among those listed.

9.2.2.2 Formulation for Flame Retardancy

Unlike the polymers mentioned in the preceding section, the remainder of the thermoplastic resins commonly used as base polymers in reinforced thermoplastics (*viz.* the polyolefins, nylons, styrenics and thermoplastic polyesters) are not inherently flame retardant. A considerable amount of work has been expended on formulations designed to improve the resistance of these polymers to burning through incorporation of special flame retardant additives.[8] These can enable reinforced thermoplastics based on these particular polymers to meet stringent specifications. Some materials, however, are particularly difficult to modify sufficiently: ABS is such a material, and no self-extinguishing grade of acetal polymer has also yet been made. Special additives may also be incorporated in the inherently flame retardant polymers to improve further or modify the properties. An example is the Thermocomp polysulphone Grade GF 1006 FR (*cf.* Tables 9.2 and 9.3).

Generally speaking there are two draw-backs in formulating flame retardant reinforced thermoplastics with the aid of special additives. Such additives are normally more expensive than either the reinforcement or the base polymer, and some can be very expensive, so that the cost of the modified flame retardant compound can become higher. Flame retardant additives can also affect both the moulding properties and/or physical properties of the reinforced thermoplastics which contain them. It is clear, therefore, that inherently flame retardant polymers enjoy an advantage in this respect.

Adverse effects of flame retardant additives on the properties of reinforced thermoplastics, where they occur, may be minimised by suitable choice of the additive and good formulation. The data in Table 9.3 illustrate some of the effects in the commercial Thermocomp materials of Table 9.2.

The processing effects of flame retardant additives are not invariably adverse. Thus the flame retardant additives can improve the flow properties and increase the flow length, facilitating the moulding of thin-walled sections and promoting a good surface finish. Long dwell times in the cylinder and high melt temperatures can promote decomposition of some halogenated additives (see below) which in turn may result in gassing and in possible corrosion of cylinder and mould surfaces. Chrome plating of mould surfaces is recommended for this reason. It has also been pointed out[9] that the improved flow properties of many reinforced thermoplastics and compounds containing flame retardant additives make it possible to reduce the melt temperatures in moulding thus minimising the potential difficulties just mentioned.

9.2.2.3 Flame Retardant Additives

The additives employed in commercial practice are often chlorinated and/or brominated compounds or their combinations with antimony trioxide ('antimony oxide'). Typically, the halogenated compounds include hexabromobenzene, and chlorinated ethers (e.g. decabromodiphenyl ether) as well as a number of proprietary materials whose chemical nature is not fully disclosed. A typical ratio for the combination of a halogenated compound with antimony oxide would be about 2·5:1. Exact formulations are closely guarded commercial secrets of polymer suppliers and independent compounders. Flame retardant additives are mentioned in the Patent Specifications summarised in Appendix 1.

The level of addition of flame retardants varies somewhat with the nature of the additive and the reinforced polymer system. For glass-fibre-reinforced nylon which, because of its importance as an engineering plastic and the associated high volume of use, it is particularly important to obtain in flame retardant form, the level of additive content (with a typical common additive system) may have to be between 20 and 30% if V-0 rating is to be achieved. Such comparatively high additive content may adversely affect mouldability, tensile strength and impact strength: i.e. the effect may be similar to that of a heavy loading with non-reinforcing filler. Proper formulation, and the use of suitable additives, can minimise these effects and formulations are possible which maintain the high level

Some Properties of Flame Retardant Versions of Commercial Reinforced Thermoplastics (Thermocomp Materials)
(Courtesy of Liquid Nitrogen Processing Corporation)

Base polymer	Glass content (%)	LNP series code	Physical: Specific gravity	Physical: Water absorption 24 h	Physical: Mould shrinkage ($\frac{1}{8}$ in section)	Mechanical: Tensile yield strength	Mechanical: Flexural strength	Mechanical: Flexural modulus	Mechanical: Izod impact strength notched/unnotched	Mechanical: Flexural creep at 73°F 2 000 psi at 1 000 h	Thermal: Deflection temp. at 264 psi	Thermal: Coefficient of thermal expansion
	Units ASTM		D792	% D570	in in^{-1} D955	psi D638	psi D790	psi D790	ft lb in^{-2} D256	% D674	°F D648	10^{-5} in in^{-1} °F^{-1} D696
SAN	30	BF-1006 FR	1·50	0·05	0·001	14 000	20 000	1 600 000	1·0/3–4	0·16	190	2·1
Polycarbonate	30	DF-1006	1·43	0·07	0·002	18 500	28 000	1 200 000	3·7/17–18	0·20	300	1·3
Polycarbonate	10	DFA-113	1·25	0·11	0·003	9 500	16 000	500 000	2·3/30–40[a]	0·42	290	1·8
Polycarbonate	30	DFL-4036[b]	1·55	0·06	0·001 5	17 500	26 000	1 200 000	2·0/10–12	0·20	290	1·5
Polysulphone	30	GF-1006	1·45	0·20	0·002 5	18 000	24 000	1 200 000	1·8/14–15	0·18	365	1·4
Polysulphone	30	GF-1006 FR[c]	1·46	0·20	0·002 5	17 500	24 000	1 200 000	1·3/12–13	0·20	365	1·4
PPO	30	NF-1006	1·27	0·06	0·001	21 000	27 000	1 100 000	2·0/12–13	0·20	360	1·5
Polypropylene	20	MF-1004 FR	1·31	0·02	0·003	6 000	11 000	750 000	1·0/1–2	0·62	265	2·4
PVC (Copolymer)	15	VF-1003	1·54	0·01	0·001	13 000	18 000	850 000	1·2/6–7	0·22	155	1·8
Nylon 6	30	PF-1006 FR	1·58	0·85	0·003	22 000	33 000	1 350 000	1·5/14–16	0·32	4·5	1·8
Nylon 6.10	30	QF-1006 FR	1·58	0·22	0·004	18 500	30 000	1 200 000	1·5/12–14	0·33	400	1·8
Nylon 6.6	30	RF-1006 FR	1·62	0·60	0·004	22 500	33 000	1 450 000	1·3/10–11	0·29	460	1·8
Polyester (PTMT)	30	WF-1006 FR	1·69	0·06	0·002 5	17 500	24 000	1 400 000	1·2/6–8	0·24	400	1·6
Modified PPO	20	ZF-1004 FR	1·23	0·06	0·002	15 000	19 000	850 000	2·0/12–14	0·25	275	1·9

FR = Special flame retardant formulation.

[a] $\frac{1}{8}$ in test bar.

[b] DFL-4036—30% glass reinforced, 15% TFE lubricated polycarbonate.

[c] Low smoke generation grade.

of physical properties and good processability whilst meeting meaningful flame retardant specifications.

9.3 LOW-FRICTION MATERIALS

The reinforced thermoplastics which fall under this heading are those with good (low) friction and frictional wear properties. Such materials, sometimes also called 'self-lubricating', are of interest for use in bearings, bearing retainers, bushings, gears, cams, slides and other working parts where the combination of their special properties with the improvement in mechanical and thermal ones conferred by the reinforcement, is particularly valuable.

9.3.1 Types of Low-Friction, Reinforced Thermoplastic Materials

Among the 'engineering polymers', nylons, acetal polymers and thermoplastic polyesters (PBT) have comparatively low friction coefficients and good wear properties and are often used, unfilled, for the applications just mentioned. Nylon 6.6 or 6 and polyacetal are the most common base polymers in low-friction reinforced thermoplastics. The incorporation of reinforcement raises the friction coefficient and wear propensity. Addition of certain 'internal lubricants' can counteract this effect and in fact improve the friction and wear properties of the compound beyond those of the base polymer alone. Similar results can be obtained in reinforced thermoplastics based on other polymers.[10] The internal lubricants in commercial use are:

(a) PTFE particles (fibres have also been used in a polyacetal—Delrin AF);
(b) PTFE and a silicone;
(c) graphite: carbon fibres can also act as an internal lubricant, whilst imparting strength and hence improving the potential of the part to operate under high load.[11] However, the basic carbon fibre types differ somewhat in their lubricant effects; type I (high modulus) fibres have been found generally more suitable;[12]
(d) molybdenum disulphide: this is used, normally in proportions of a few per cent, mainly in reinforced compositions based on nylon 6.6 or nylon 6. The effect on the coefficient of friction is not very great, but the resistance to frictional wear is considerably improved.[13]

The silicones which are used as secondary lubricants with PTFE have also been described as 'migratory lubricants'[14] because they are believed to migrate to the wear surface during operation. They are also claimed to eliminate the initial run-in period which occurs with other low-friction

TABLE 9.4

Wear Friction and Performance Properties of Some Thermoplastics with and without Reinforcement and/or Internal Lubricants

(Data of Theberge *et al.*[14] reproduced, with permission, from *Machine Design*, The Penton Publishing Co.)

Base resin	Lubricant (%)			Wear factor (K^a)	Coefficient of friction		Limiting *PV*		
	Silicone	PTFE	PTFE and silicone		Static	Dynamic	at 10 ft min^{-1}	at 100 ft min^{-1}	at 1 000 ft min^{-1}
Acetal copolymer	—	—	—	65	0·14	0·21	4 000	3 500	<2 500
	2	—	—	29	0·09	0·12	4 000	5 000	3 000
	—	20	—	17	0·07	0·15	10 000	12 500	5 500
	—	—	20	9	0·06	0·11	8 000	15 000	12 000
Nylon 6.6	—	—	—	200	0·20	0·28	3 000	2 500	>2 500
	2	—	—	40	0·09	0·09	3 000	6 000	9 000
	—	20	—	12	0·10	0·18	14 000	27 500	8 000
	—	—	20	6	0·06	0·08	14 000	30 000	12 000
Nylon 6.6 (30% glass)	—	15	—	16	0·19	0·26	17 500	20 000	17 000
Nylon 6.10 (30% glass)	—	15	—	15	0·23	0·31	20 000	15 000	12 000
	—	—	15	8	0·17	0·18	20 000	15 000	12 000
Polycarbonate	—	—	—	2 500	0·31	0·38	750	500	—
	—	15	—	75	0·09	0·15	15 000	20 000	10 500
	—	—	15	40	0·08	0·09	15 000	22 500	17 000
Polystyrene	—	—	—	3 000	0·28	0·32	750	1 500	—
	2	—	—	37	0·06	0·08	4 000	9 000	—

[a] A wear factor of $K = 1$ indicates a test condition that produces a wear volume of 1 in^3 of material in 1 h at a load of 1 psi and a velocity of 1 ft min^{-1}. Wear data presented here (to be multiplied by 10^{-10}) were determined on a 0·900-in ID × 0·125-in OD plastic thrust washer operating at 50 ft min^{-1} under a 40-psi load, against a 1 040 steel washer with a 12-rms finish.

materials before a film of the lubricant is formed on the co-working surface.

Since, in an internally lubricated reinforced thermoplastic material, the presence of reinforcement usually tends to counteract the effects of the lubricant, there is, for each type of internally lubricated reinforced thermoplastic, an optimum value or range for the respective contents of reinforcement and lubricant, and the ratio between them. As pointed out by Theberge, the main factor limiting the lubricant content in a lubricated reinforced thermoplastic is the effect on structural integrity,[10] and—more generally—on properties, and especially mechanical properties.

The frictional and wear properties of some polymers and the effect thereon of internal lubricants are illustrated in Table 9.4.

9.3.2 Assessment of Performance of Lubricated Reinforced Thermoplastics

Frictional and wear properties of these compounds can be measured by the methods already mentioned in Chapter 6 (Section 6.2.2.2). The value of the greatest interest in connection with performance in practice (in bearings) is the limiting PV value to which reference has also been made. This is the highest value of the product of specific bearing load and peripheral velocity consistent with satisfactory operation over long periods. It is related to the conditions at the interface (in particular local temperature and friction) and is also dependent on the wear factor. These relationships are used as bases for test methods whereby PV limits for materials are established.[15] The test apparatus is essentially a bearing run under controlled, dry conditions, in which the polymer works in contact with a metal surface. The data from such tests can also be used to determine the optimum loading. For PTFE as the lubricant this is approximately 15% in glass-reinforced nylon and acetal polymers.

REFERENCES

1. Fennimore, C. P. and Martin, F. J. (1966). *Modern Plastics,* **44,** 3, 141.
2. Abbott, C. (1973). *Europlastics,* **46,** 11, 68.
3. British Standard 3456: Section 2.10:1972: Specification for the Testing and Approval of Household Electrical Appliances; Part 2 Particular Requirements; Section 2.10 Room Heating and Similar Appliances.
4. Schwarz, K. H. (1962). *Electrotech. Z. (ETZ),* B **14,** 14th May, 273 and 291. British Electrical and Allied Industries Research Association Translation Ref. LB 2072.
5. Scherbaum, R. (1972). *Electrotech. Z. (ETZ),* B **24,** 4, M29.
6. Reymers, H. (1970). *Modern Plastics,* October, 92.

7. Theberge, J. E., Arkles, B. and Cloud, P. J. (1971). 26th Conference RP/C Division, SPI, Paper 3-A.
8. Theberge, J. E., Arkles, B. and Cloud, B. J. (1971). *Machine Design,* 4th February and 30th September.
9. LNP Flame Retardant Fortified Polymers, LNP Bulletin 260-573, May 1973.
10. Theberge, J. E. (1970). 'Properties of Internally Lubricated Glass-Fortified Thermoplastics for Gears and Bearings', Plastics Institute Conference on Reinforced Thermoplastics, Solihull, England, October, Paper 14.
11. Anon. (1973). *Modern Plastics International,* December, 62.
12. Perkins, C. W. (1970). 'The Tribological Characteristics of Carbon-fibre Reinforced Thermoplastics', Plastics Institute Conference on Reinforced Thermoplastics, Solihull, England, October, Paper 12.
13. Reichelt, W. (1970). 'Properties, Injection Moulding and Applications of Reinforced Nylon', *ibid.,* Paper 6.
14. Theberge, J. E., Arkles, B. and Goodhue, R. E. (1953). *Machine Design,* 27th December, 58.
15. Liquid Nitrogen Processing Corp. Bulletin 610, 1970.

CHAPTER 10

REINFORCED FOAMS

10.1 NATURE AND DEFINITION

Reinforced thermoplastic foams are normally encountered in the form of moulded articles. They may be regarded as a group within the wider, general class of structural foams in that they are structural foams containing reinforcing filler, usually fibrous (and commonly glass fibre).

Structural foam is a cellular material with a solid integral layer ('skin') at the surface and the content and size of voids (foam cells) increasing towards the centre of the section. The term is usually applied to products of this nature manufactured by injection-moulding. This is in accord with the origin of the name, which was first applied by the Union Carbide Corporation to foamed mouldings produced by their patented injection-moulding process (see Section 10.4.1). Extruded foamed sheet with an integral skin is also occasionally referred to as structural foam, as is the integral-skin variety of rigid polyurethane foams. However such foams are thermosetting materials and will not, therefore, be described here.

10.2 CHARACTERISTICS AND FORMATION—GENERAL

From the point of view of processing, the structural foams' principal *raison d'être* is that they make possible the moulding of large components of acceptable strength and stiffness on equipment which does not require the high clamping forces necessary in conventional injection-moulding of large solid components. Much of the cost of large injection-moulding machines is accounted for by the equipment required to keep the mould halves closed under the extremely high injection pressures. A structural foam part can be produced at pressures of about one-tenth of those used for moulding comparable solid components. A further factor in the economy of production is the fact that relatively cheaper tools may be employed than in the production of solid mouldings. In most commercial moulding of structural foam the pressures in the mould are comparatively low (*cf.* Section 10.4). Hence beryllium–copper,* kirksite and aluminium (steel-framed) moulds are suitable instead of the normally used tool steel moulds. In addition to these advantages in processing costs, structural foam mouldings also offer a material cost saving: because of the voids content the moulding of a given volume will contain less polymer than a solid moulding, whilst the stiffness and strength properties, especially in fibre-reinforced structural foams, are good. Indeed, one of the advantages of reinforced structural foams, and to a lesser extent structural foams generally, is their high stiffness-to-weight and flexural strength-to-weight ratio in comparison not only with 'engineering' plastics but also with metals.

Theoretical treatment suggests that the modulus of a foamed material can be calculated on the basis of a simple theory of mixtures from the modulus of the polymer (or the base polymer/reinforcement composition in the case of reinforced foam) and that of air which is the second component of the mixture,[1,2] and that in this system it will thus be a function of the percentage of voids. This is confirmed in practice, but in structural foam there are two additional factors which increase the stiffness of the component over that predicted on the basis of this type of simple calculation. These factors are the increase of density from the interior to the surface (which has an additional structural rigidifying effect in a complete moulded part, and also makes for increased flexural strength), and the orientation of reinforcing fibres within the foam cell walls. This factor is further discussed below.

As well as the structural and processing advantages just mentioned, structural foam also offers the following product advantages.

Relative freedom from moulding stresses

This is possible because, in the course of production (*cf.* Section 10.4), the mould is filled by the internal expansion of material, minimising such sources of internal stress in ordinary injection-mouldings as orientation stresses and packing stresses.

Freedom from sink marks

Mouldings with variable section thickness, as well as generally thick sections, can be produced in reinforced foam with little risk of sink marks. The reason is again positive expansion of the material in the mould and the consequent complete filling of the mould.

Acceptance of nails and self-tapping screws

Properly moulded reinforced structural foam components will accept

* This may have to be chromium-plated for protection against gases released by blowing agents (see Section 10.4).

self-tapping screws and nails without the creation of disruptive stress concentrations.

Thermal insulation

In many cases the insulation properties of a structural foam moulding are superior to those of an analogous solid moulding. This may be of value, for example, in trays which are frequently produced in structural foam.

Sound damping

Structural foam parts have this effect which is beneficial in acoustic applications, e.g. gramophone cases which have been produced in reinforced structural foam.

Two disadvantages of structural foams are the longer cycle time which is normally required because the wall sections are thicker (although at comparable thickness the cycle time is similar for solid and foamed mouldings), and the tendency for swirl patterns to appear on the surface of many injection-mouldings, especially those produced by the low-pressure Union Carbide type of process (*cf.* Section 10.4.1). Where the appearance of a moulding is of importance, surface coating may have to be resorted to. Unreinforced structural foams are also inferior to the parent polymers in tensile strength and creep properties, and their heat distortion temperatures are lower.

The density of unreinforced structural foams is in the upper part of the density range for cellular plastics materials generally. Typical densities are 40–50 lb ft^{-3} (0·65–0·80 g cm^{-3}) but some mouldings are produced in densities down to 25 lb ft^{-3} (0·4 g cm^{-3}). At present the reinforcement commonly used in structural foams is glass fibre, in proportions between 10 and 30% by weight. For the same voids content, the inclusion of this high density material increases somewhat the numerical values for overall foam densities in comparison with unfilled structural foam. As has been mentioned a density gradient exists within the moulding, from a solid 'skin' on the outside, which has the density of the parent plastic, down to the lowest density in the centre of the cellular core where the cells are largest. This structure is the result of foaming in the mould, or expansion at the extruder die in sheet extrusion.

The best way to visualise the mechanism and sequence of formation of this kind of structure is to consider the factors operative during the foaming and solidification of the part in the mould. The processes have been well summarised by Wilson[2] who correctly attributes the formation of the above-mentioned density gradient basically to what he describes as 'thermodynamic properties' of the material, i.e. the way in which the cooling processes and associated viscosity changes give rise to the final structure. The cell structure is established and stabilised as the thermoplastic material cools down in the mould, whilst the gas originally dispersed or dissolved in the melt separates out and forms bubbles which become trapped to form cells within the cooling polymer. The rate of cooling is highest where the material is in contact with the mould walls and lowest in the interior which is insulated by the intervening partly molten cellular plastic. Additionally, the material flowing in contact with the mould walls undergoes the greatest amount of shear. Wilson points out that it is the combination of this shear with the most rapid cooling at the mould walls which causes the bubbles to collapse there before they reach appreciable size, displacing the gas towards the interior or out of the mould vents so that a 'solid' layer is formed at the plastic/mould interface. At the same time the closer to the centre the longer the cooling time available to the bubbles and the lower (comparatively speaking) the disruptive shear on the material. Very large cells are of course undesirable and the voids in the centre of the moulding should, therefore, not be too large. Ideally the density gradient from the solid skin on the outside to the centre of the section should be uniform and not too steep, with a corresponding uniform transition in cell size.

10.3 PROPERTIES

In terms of stiffness structural foam compares favourably per unit weight with the solid parent polymer. The same applies to such comparisons related to the wall thickness of a moulding. As has been mentioned, both the stiffness and the flexural strength of structural foam are superior to those of certain metals in the weight-related comparisons. However in a direct comparison, i.e. when standard specimens (of identical dimensions) are tested, the mechanical property values for structural foam are lower than for the solid polymer. This is illustrated by the data of Table 10.1 for polycarbonate and phenylene-oxide-based resin.

At the same time the stiffness and strength (not only flexural, but also tensile) of structural foams reinforced with even comparatively low percentages of glass fibre can equal or even exceed these properties of the solid (unreinforced) parent polymers, despite the fact that the foam may contain up to about 30% voids. In Table 10.2 this point is illustrated by figures for reinforced structural foams in polypropylene and nylon 6 in comparison with the data for these polymers unfoamed, with and without reinforcement. The figures also demonstrate the important fact that the heat deflection under load and the extensibility of reinforced structural foam are closely comparable with those of the *reinforced* parent polymer and that therefore reinforced structural foams can be greatly superior to the solid, unreinforced polymer in these respects.

Incorporation of potassium titanate whiskers (Fybex: DuPont) in

TABLE 10.1

Strength and Stiffness of Polycarbonate and Phenylene-oxide-based Resin Structural Foams and Standard Moulding Compounds

(Based in part on data abstracted, with permission, from the technical literature of the General Electric Co.)

Material	Specific gravity	Strength (psi)			Flexural modulus (psi)	Flexural strength to weight ratio	Flexural modulus to weight ratio
		Flexural	Compressive	Tensile			
PPO							
Solid (Noryl 731)	1·06	13 500	16 400	9 600	360 000	—	—
Foamed (Noryl FN-215)	0·80	6 000	5 500	3 300	240 000	100	100
Polycarbonate							
Solid (Lexan 500)	1·25	15 000	14 000	8 000	500 000	—	—
Foamed (Lexan FL-900)	0·80	10 000	7 500	5 500	300 000	133	110
Metals							
Steel (stainless)	7·90	33 000	30 000	33 000[a]	28 000 000	20	45
Aluminium (wrought)	2·80	20 000	20 000	20 000[a]	10 000 000	45	90

[a] At yield.

TABLE 10.2

Properties of Glass-fibre-reinforced Structural Foam (Chemically Blown) in Comparison with Solid Polymer and Unreinforced Foam: Nylon 6 and Polypropylene[3]

(Courtesy of Liquid Nitrogen Processing Corp.)

	Nylon 6				Polypropylene			
Composition								
Polymer (% by weight)	100	85	85	85	100	80	80	80
Glass fibre (% by weight)	0	15	15	15	0	20	20	20
Physical Properties								
Voids content (%)	0	0	10	30	0	0	10	30
Specific gravity	1·14	1·20	1·08	0·84	0·905	1·04	0·92	0·73
Mechanical Properties								
Tensile strength (ultimate) (psi) (ASTM D-638)	11 800	16 000	12 000	10 500	5 000	9 100	7 600	3 300
Elongation (ultimate) (%) (ASTM D-638)	80	2–3	2–3	2–3	200–700	2–3	2–3	2–3
Flexural strength (psi) (ASTM D-790)	15 000	23 000	18 000	16 500	7 000	11 000	8 800	6 300
Flexural modulus (psi) (ASTM D-790)	400 000	750 000	700 000	650 000	250 000	600 000	450 000	400 000
Izod impact strength ($\frac{1}{4}$ in) notched/unnotched (ft lb in^{-2}) (ASTM D-256)	1·0/no break	1·2/7–8	0·6/5–6	0·4/3–4	0·5–2/—	1·4/6–7	1·0/4–5	0·9/3–4
Deformation temperature at 264 psi (°C) (ASTM D-648)	70	202	199	199	60	140	94	73

polystyrene foam increases the stiffness, strength and dimensional stability of the foam, and promotes finer cell structure through a nucleating effect.

Deformation under load of glass-reinforced polystyrene and nylon structural foams based on materials marketed by the Fiberfil Division of Dart Industries Inc., is illustrated in Fig. 10.1.

In structural foam, the reinforcing effect of the fibrous filler arises essentially in the same way as it does in unfoamed mouldings. The presence of the reinforcement improves considerably the strength and stiffness as well as other properties of the solid material (the cell walls) of the foam. The effects are further enhanced by the tendency of the reinforcing fibres to become oriented in the course of cell formation and solidification of the cell wall material. Finally, superimposed on these local effects is the overall strengthening effect of the reinforced structural skin in which the reinforcing fibres may also be oriented. As pointed out by Wilson[2] in a foaming, glass-fibre-bearing thermoplastic melt in the mould the glass fibres tend to align themselves within and along the cell walls when the cells are being formed (Fig. 10.2). This local anisotropic effect increases the strength and stiffness of the cell walls.

In so far as in a properly foamed moulding the cells are roughly

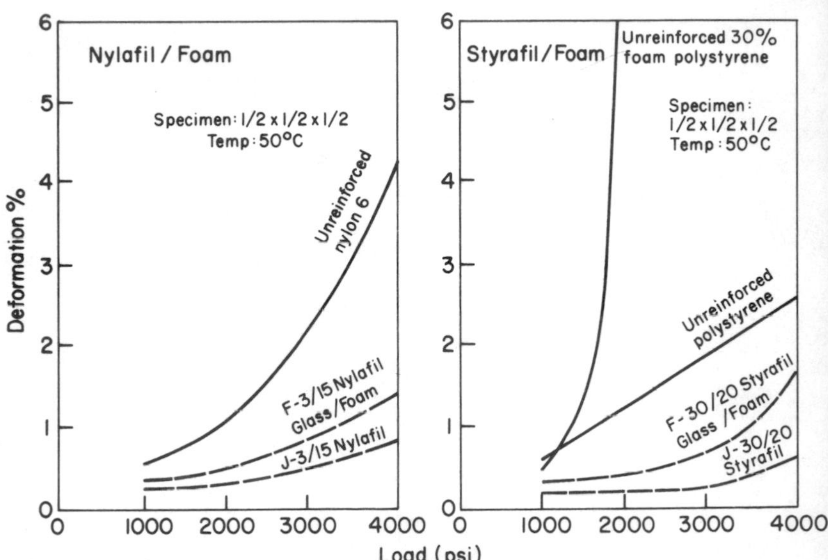

FIG. 10.1 Deformation under load of glass-fibre reinforced structural foam. Nylafil[R] F3/15 = Fiberfil nylon with 15% glass fibre (*cf.* Chap. 2, Table 2.4). Styrafil F30/20 = Fiberfil/polystyrene with 20% glass fibre. (Reproduced, with permission, from the Technical Bulletin *Glass Reinforced Thermoplastic Foam Resins* of Fiberfil.)

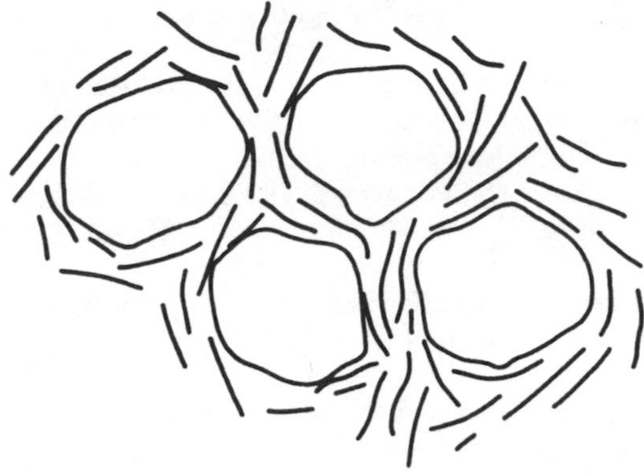

FIG. 10.2 Orientation of glass fibres in the cell walls, schematic representation.

spherical, the cell walls themselves have no particular directional alignment, and may be regarded as randomly oriented overall. Thus the local orientation and anisotropic reinforcement effect of the fibres within individual cell walls are integrated into, and become an additional factor in, a general enhanced reinforcing effect which is isotropic in the moulding considered as a whole. It is interesting to note that this effect can be realised to the full because the foaming in the mould counteracts the tendency, common in 'solid' reinforced thermoplastic mouldings, for preferential flow-induced orientation of the fibrous filler.

10.4 PRODUCTION: INJECTION-MOULDED STRUCTURAL FOAM

Certain differences exist between the United States and Europe in the techniques and practice of production of structural foams generally, including the reinforced variety. Since virtually all of the reinforced structural foam production is accounted for by injection-moulded articles, the discussion here will be confined to injection-moulding processes.

In the United States the evolution of manufacturing techniques has been largely associated with, and based on, proprietary processes. In particular the Union Carbide process involving direct injection of nitrogen into the polymer melt has formed the basis of much of the current manufacturing practice. This type of process utilises highly modified injection equipment. The European development has relied upon the use of conventional injection-moulding machinery with comparatively minor modifications. In European practice foaming is normally achieved by 'chemical' means, through decomposition of blowing agents incorporated in the moulding compound.

Foam moulding methods may be broadly divided into 'low pressure' and 'high pressure' types, depending on whether or not the pressure in the mould is substantially lower than that on material being injected. Of the commercially important processes outlined in Sections 10.4.1–10.4.3 only the USM process is a high-pressure method.

10.4.1 Entrained-gas Blowing (The Union Carbide Process—predominantly in the USA)

At the present time (1974) there are over 30 operators of this process in the United States, this number having been built up since the process made its first appearance in 1965. Its principle is the introduction of nitrogen under pressure into the polymer melt and injection of the pressurised, gas-containing melt into a mould held at much lower pressure, in quantity sufficient to fill the mould after the correct volume increase through expansion has taken place.[4] The equipment consists of an injection machine with a continuous-extrusion unit feeding an accumulator. Nitrogen under pressure is injected through a vent into the polymer melt in the extruder (feed section of the screw). The melt containing the gas in fine dispersion or solution is passed into the accumulator, where it is held under a pressure of up to 5000 psi. When the required charge has built up in the accumulator it is injected by a plunger into the mould. This is under low pressure (about 300 psi) and the charge expands to fill it, the volume of the shot being so adjusted as to give the required density of the finished moulding. Cooling and extraction of the moulding then proceed in the conventional way. Clearly a major advantage of the process is that unmodified feed stock can be employed, i.e. no addition of chemical blowing agent is required. This can give substantial economies in the material cost. However, it has the disadvantage of higher equipment capital cost in comparison with the conventional injection route. Moreover, since it is a patented proprietary process, licence fees are incurred. The process is schematically illustrated in Fig. 10.3.

The Union Carbide process is as yet little used in Europe, although its wider adoption is being confidently predicted.[5]

10.4.2 The USM Process

This process was developed in the USA by the United Shoe Machinery Company. It utilises conventional 'chemical' blowing, i.e. the decomposition within the polymer melt of a chemical additive (the blowing agent) which releases a large volume of gas (most commonly nitrogen) ultimately responsible for the blowing. It is widely accepted[6] that in this

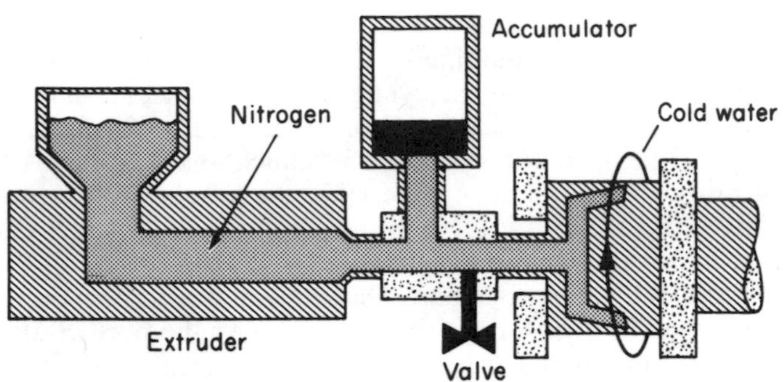

FIG. 10.3 Schematic representation of the entrained-gas blowing process. (Reproduced, with permission, from Technical Publication SFR-1 (30M) 4-73 *Structural Foam Resins* of the General Electric Co.)

type of system the gas is in fine dispersion, possibly solution, in the polymer melt until the pressure on the latter is reduced (within the mould —see below) so that the gas can separate out and expand to form the required cellular structure.

The method differs radically from the Union Carbide process also in that in the former the polymer melt virtually expands into the mould, whilst in the USM method it initially remains at a high pressure requiring considerable locking forces to hold the mould halves together during the initial stage of injection.

The basic equipment used in the USM process consists of a fairly conventional injection-moulding set-up capable of high locking pressures on the mould (moulding pressures 10 000–20 000 psi), but special moulds with an expandable cavity (e.g. by means of a movable base plate[7]) are required. A full-size shot is introduced into the mould under the appropriate (high) pressure; the mould cavity is then expanded so that the charge can in turn expand to fill the enlarged space inside the mould. The special expanding moulds are expensive and subject to shape limitations.

10.4.3 The Reduced-Shot Method

10.4.3.1 PRINCIPLE, PROCESS AND EQUIPMENT

In this method the blowing gas is generated chemically, i.e. by the decomposition of an admixed blowing agent in the hot polymer melt. A 'short shot' (quantity of polymer melt less than that required to fill the mould) is injected, so that expansion proceeds under lowered pressure in a fixed mould. The mould may be kept cold to promote skin formation.

These features make it possible to use conventional injection-moulding equipment, only slightly modified. The modifications would include provision of a short nozzle for the cylinder equipped with a shut-off valve operated pneumatically, mechanically or hydraulically. The latter feature is probably the most important single modification. The reduced valve length cuts down the amount of material in the nozzle and its dwell time there, whilst the nozzle shut-off prevents drooling of the melt from the cylinder in which the melt is pressurised by the gas generated through decomposition of the blowing agent. The screw may also be modified and the heating adjusted to ensure proper control over the plasticisation and gasification of the material in the cylinder. As injection speed must be extremely fast it is recommended that the following techniques be used to increase the speed of filling the tool:

(a) the use of booster circuits or oil diverted from the clamp;
(b) employing over-size pumps on the injection machine;
(c) the use of a pressure switch to delay forward movement of the screw and produce a forceful explosive filling of the mould.

In Europe over 90% of foam moulders are using the reduced-shot technique,[5] which offers the additional advantage of freedom from patent restrictions.

10.4.3.2 Blowing Agents

The blowing agent may be dry-blended with the moulding compound, in proportions within the range 0·15–1·5% by weight of the polymer. This is normally adequate to produce voids contents up to the desirable limits. For example the highest voids contents shown in Table 10.2 (30%) were obtained with about 0·5–1·0% blowing agent. Mouldings with thin-wall sections need proportionately more blowing agent (for the same ultimate density) than thick-walled mouldings. Apart from unnecessary expense, the use of too much blowing agent can cause uneven blowing, extend the cycle times required, and even promote degradation of the polymer in extreme cases.

The addition of 'a pinch' of chemical blowing agent—to initiate cell nucleation—has been suggested[3] as a way of counteracting the formation of large, irregular cells which can occur in the entrained-gas (Union Carbide) blowing method.

The main requirements of a blowing agent for use in moulding structural foam are:

(a) compatibility with the polymer melt: the blowing agent should be easily dispersed or dissolved in the melt and should not affect the polymer chemically even at the highest melt temperatures;
(b) gas release at the correct rate within the appropriate temperature range;

(c) non-toxicity and non-aggressive nature of the agent itself and of the decomposition products (blowing gases, by-products and residues).

Stability at normal temperatures and long shelf life are also desirable.

The following blowing agents are of particular interest in connection with 'chemical' blowing of injection-moulded structural foam.

Azodicarbonamide (*ADA*), sometimes called azobisformamide (ABFA). This is a non-inflammable powder decomposing at 170–220°C: nitrogen, carbon dioxide and carbon monoxide are the main gaseous decomposition products, evolved in very high yield. Suppliers include: Fisons Industrial Chemicals (USA: *Ficel AC*; UK: *Genitron AC*), Bayer (Germany and USA: *Porofor ADC*), National Polychemicals Inc. (USA: *Kempore*), Fairmond Chemical Co. (USA: *Axocel*).

5-Phenyltetrazole (*5PT*). This solid (powder) blowing agent decomposes at polymer melt temperature liberating nitrogen as the blowing gas and leaving behind phenolic residues. Supplier: National Polychemicals Inc. (USA: *Expandex OX 5PT*).

p-toluenesulphonylsemicarbazide (*TSS*). This blowing agent is a powder decomposing within the range 190–270°C. The blowing gas liberated is nitrogen. Supplied by Uniroyal, Chemical Division (USA: *Celogen RA*).

Trihydrazine triazine (*THT*). Another pulverulent blowing agent, operating within the range 240–300°C, to liberate nitrogen as the blowing gas. Supplied by National Polychemicals Inc. (USA: *Expandex THT*) and Fisons Industrial Chemicals (USA: *Ficel THT*).

Whilst these reagents are, generally speaking, suitable for blending with most reinforced thermoplastic moulding compounds, only 5PT is recommended for use with polycarbonate—a polymer sensitive to degradation by some of the products of decomposition of nitrogenous blowing agents. However, the coloured residues which can be generated by 5PT make this reagent unpopular with some users.

Apart from dry-blending the powders directly with the moulding compounds, blowing agents may be incorporated also through blending with concentrates in what is essentially a masterbatch technique. Glassfibre-reinforced thermoplastics concentrates typically containing about 20% by weight of a blowing agent are available, e.g. from LNP and its licensees (concentrates, PDX series). Foaming glass-reinforced compounds are also available containing blowing agents (and in many cases nucleating additives) pre-compounded at normal concentration levels (e.g. the Fiberfil compounds whose deformation properties are shown in Fig. 10.1).

To date reinforced structural foams have been produced (not all on a large, commercial scale) from the following polymers (several available as foamable compounds or concentrates from main suppliers): polystyrene, ABS, polypropylene, HD polyethylene, nylon, modified PPO, a modified vinyl polymer (Zerefil/Foam F200/20: Fiberfil), polycarbonate, polysulphone, thermoplastic polyester, ETFE copolymer (Tefzel).

10.4.4 Other Methods

Several other methods have been announced from time to time, without however great immediate impact on commercial practice. Most are variants of those described in the preceding sections: two may be briefly mentioned by way of illustration.

10.4.4.1 The ICI Twin-injection Process

This process patented by ICI for injection-moulding structural foam involves separate synchronised injection of normal moulding compound to form the solid skin of the moulding, and of the foaming material to form the cellular core.

10.4.4.2 The Allied Thermoplastic Foam Process[8]

This process employs a special manifold, integral with the mould. The manifold is initially pressurised by an inert gas, but is subsequently opened to permit expansion after the polymer melt, containing a blowing agent, has been injected into the mould.

10.4.5 Moulding Faults

The main faults which may arise in structural foam moulding are listed in Table 10.3, together with the most likely causes. The list as originally compiled (by the General Electric Company[9]) related to unreinforced structural foam (with special reference to polycarbonate and modified PPO) but it is largely applicable also to reinforced structural foam based not only on these but also on other polymers.

10.5 APPLICATIONS

The commercial development of reinforced structural foam applications has been fairly slow. Some uses, originally predicted, and now to some extent initiated, include furniture panels and components (glass-reinforced polystyrene foam), electrical components and minor parts for motor cars (glass-reinforced nylon), housings and components for business and office machines and appliances (glass-reinforced polycarbonate, modified PPO). Water skis have recently been moulded in glass-reinforced polycarbonate structural foam. This material has also been used to produce telephone distribution boxes, formerly made in concrete: ease of assembly is a considerable advantage in this application.[5] Perhaps the greatest progress has been made in the introduction of large foamed reinforced polyolefin mouldings.

Materials-handling containers (3 gallon capacity: weight 24 lb) were being produced in foamed, glass-fibre-reinforced HD polyethylene as

TABLE 10.3
Structural Foam Moulding Faults and Their Causes
(Reproduced, with permission, from Technical Publication SFR-1 (30M) 4-73 *Structural Foam Resins* of the General Electric Company.)

Fault	Likely cause
Moulded part	
Flash	Clamp pressure too low Shot size too great
Heavy part	Shot size too great Stock temperature too low Fill time too slow Gas pressure too low Not enough blowing agent (chemical system)
Short fill	Stock temperature too low Nozzle not fully open Shot size too small Mould not properly vented Heater band not functioning
Post-blow	Cooling cycle too short Overpacked shot High gas pressure Non-uniform manifold temperature Mould temperature too high Too much blowing agent (chemical system)
Material discoloration	Stock temperature too high and/or residence time too long Non-uniform heating resulting in hot spots
Poor weld line strength	Stock temperature too low Inadequate venting Flow length too long Injection speed too slow
Poor surface appearance	Mould not properly vented Stock temperature too low Injection speed too slow Flow length too long
Large 'glossy' voids	Mould not properly vented Stock temperature too high Gas pressure too high Flow length too long Shot size too small Non-uniform melt temperature

Table 10.3—*continued*

Fault	Likely cause
Process	
Slow injection speed	Stock temperature too low Nozzle not fully open Poor temperature balance Injection gas pressure too low
High extruder amperage	Stock temperature too low
Nozzle freezing	Mould hung on nozzle openings Nozzle temperature too low Improper nozzle shut off

early as 1971.[10] More recently (1974) a fruit crate weighing 32 kg has been produced in Europe in the same material (20% by weight glass fibre) with two injection machines feeding a single mould: despite the very large area the locking force required was only about 800 tons.

Other applications of structural foam are also mentioned in Chapter 8. An interesting development has been the use of expanded, glass-fibre-reinforced ABS for motor car seat-back components (Chevrolet and other cars).[11]

REFERENCES

1. Baxter, S. and Jones T. T. (1972). *Plastics and Polymers,* **40,** 146, 69.
2. Wilson, M. G. (1971). *SPE Journal,* June, **27,** 35.
3. Theberge, J. E. and Cloud, P. J. October, 1973: LNP Notes on Structural Foam Moulding.
4. Cochran, L. D. and Osborn, C. W. (1969). *SPE Journal,* September, **25,** 20.
5. Miller, W. G. (1973). *Europlastics,* **46,** 11, 62.
6. Titow, W. V. and Lanham, B. J. (1971). *Penn's PVC Technology,* Applied Science Publishers, London, 481.
7. Challenger, P. (1970). *Polymer Age,* **1,** 9.
8. Weir, L. C. (1969). *Modern Plastics,* March, 69.
9. Structural Foam Resins: Technical Booklet SFR-1(30M) 4-73, General Electric Company, 1973.
10. Lesseliers, L. (1972). 27th Conference, RP/C Institute, SPI, paper 14-E.
11. Anon. (1972). *Modern Plastics International,* October, 53.

APPENDIX 1

REINFORCED THERMOPLASTICS IN PATENTS

The patent literature relating to various aspects of reinforced thermoplastics materials is extensive in all the principal geographical zones of manufacture. Any active and developing technology, and reinforced thermoplastics certainly fall within this category, is usually strongly reflected in patent literature. Reinforced thermoplastics are no exception.

The review in this section is not meant to be a complete patent survey. As in many other areas of technology, most of the relevant patents in one country are equivalents of those in other countries, so that the total number of patent specifications in a given field is normally much greater than the actual number of patented processes and products.

In the present section a list of United Kingdom patents in the field of reinforced thermoplastics is used to illustrate something of the 'state of the art' as reflected in patent specifications. Most of the specifications significant in this sense have been covered and the general pattern is sufficiently similar to those in the rest of Western Europe and the United States to be reasonably widely representative.

The patents listed start essentially from the early 1960s, but there are two earlier patents of particular interest granted, respectively, to DuPont and the Koppers Co. British Patent No. 618 094 to DuPont (published in February 1949, application in the USA October 19th, 1945) covers the extrusion compounding of short glass fibres into various thermoplastics. The fibres mentioned in the specification are rather shorter than those normally used at the present time for the manufacture of reinforced thermoplastics; it is an important feature of the patent's main claim that they have aspect ratios of less than 40 : 1. This disclosure of the extrusion compounding method pre-dates by several years the spate of patents dealing with extrusion compounding which issued in the early 1960s. It also similarly pre-dates the key Fiberfil US Patent (No. 2 877 501 published on March 17th, 1959, application date December 24th, 1952) for moulding materials of the long, parallel fibre type (the second main product variant of current commercial importance—*cf.* Chapter 5). It is noteworthy that, for non-technological reasons, no British equivalent of the Fiberfil US patent was ever granted, whilst a patent for parallel, long-fibre moulding products was obtained in the UK by the Koppers Co. Inc. (British Patent 791 663): the specification was published on 5th March, 1958, i.e. before the publication in the USA of the Fiberfil specification. Like the Fiberfil patent in the United States, the Koppers UK patent is a major early one for this particular type of reinforced moulding compound. However, the Koppers patent is limited in scope in comparison with the rather similar though much later specification of the Rexall Drug Company's British Patent 1 167 849 (see below and also Chapter 5, Section 5.2.3.1), in that it covers essentially only polystyrene as the base polymer and its application to the glass strands in aqueous dispersion. In the American Fiberfil specification several base polymers are named and the more commercially important melt coating process is covered, as well as the dispersion and solution coating variants.

UK Patent No.	Patentee	Comments
950 656	ICI Ltd.	Extrusion compounding process and product claims primarily relating to nylon but other thermoplastics included. Length of fibrous filler in moulding granules and finished products specified
1 010 043	Societe Organico	Product and process patent relating to intimate mixture of powdered thermoplastic fibres and non-fibrous fillers, which may be heat-compounded
1 055 395	Dow Chemical Co.	A process patent relating to an improved method of direct moulding of a mixture of fibrous filler and polymer powder or granules
1 067 940	ICI Ltd.	Epoxy resin coating on glass or other reinforcing fillers
1 073 804	Sumitomo Chemical Co. Ltd.	Natural talc-filled polypropylene composition produced by blending
1 086 980	Bayer	Glass-reinforced ABS product and process claims. Produced by roll compounding
1 087 859	Dow Chemicals	Process patent relating to the introduction of glass fibres at the vent of a compounding extruder
1 094 439	Hercules Inc.	Glass-reinforced polypropylene wherein the polypropylene is chemically modified by reaction with ethenically unsaturated

		compound, e.g. maleic anhydride, to improve coupling of the glass to the polymer. Basically a patent for coupled polypropylene system
1 095 700	ICI Ltd.	Filled compositions in which a two-component coupling agent is used
1 100 924	Bayer	Glass-filled polyamide articles. Composition apparently melt compounded by roll mixing
1 113 387	BASF	Moulding material containing a fibrous filler in conjunction with a small proportion of inorganic pigment. Smooth surface finish claimed. Apparently compounded by extrusion
1 131 533	Bayer	Introduction of coupling agent into the melt in fibre-filled polyamide compositions prepared by extrusion compounding
1 167 849	Rexall Drug Co. (now Dart Industries Inc.)	Glass-reinforced concentrate granules of the parallel long glass fibre type. Method and product: similar to the Fiberfil process
1 179 216	Ferro Corp.	A fire retardant composition comprising low or high density polyethylene with polypropylene. May be filled with glass or asbestos
1 181 425	Turner Bros. Asbestos Co. Ltd.	Propylene polymer reinforced with chrysotile asbestos containing an antioxidant system. Asbestos fibres may be pre-treated with calcium stearate or silicone
1 182 439	Chemische Werke Huls AG	Mixtures of polyolefins said to be resistant to stress cracking (high density polyethylene with propylene and butene polymers or copolymers). May be filled or reinforced
1 187 957	British Industrial Plastics Ltd.	Method of removing unstable materials from acetal polymers by heating the polymer in the presence of glass fibres, beads or asbestos fibre. The fibres may remain in the compound to provide a reinforced material
1 199 283	Turner Bros. Asbestos Co. Ltd.	Reinforced PVC sheets formed from a dough-like composition

		containing PVC with minor proportions of an acrylonitrile/butadiene copolymer or polyacrylate and toluene, talc and asbestos fibre
1 200 342	ICI Ltd.	Fibre-reinforced thermoplastic material at least initially in mat form produced by fusing lower melting fibrous component around the higher melting fibres. Product may be chopped up to form moulding compound
1 204 835	Farben Fabriken Bayer	Self-extinguishing nylon, including filled grades, but flame retardant additive being a melamine derivative (between 0·5 and 25%)
1 211 083	Monsanto	Filled polyamide (polylactam) polymers produced with filler present in the monomer solution before condensation
1 211 633	Monsanto	Polymerising a polyimide onto fibrous boron
1 213 247	Dart Industries	Asbestos (chrysotile) reinforced nylon also containing 0·5–10% low density polyethylene
1 215 481	NRDC	Laminates made from fibre or filler matrices impregnated with polyamide prepolymer or its solution
1 226 711	Toray Industries Inc.	Oxazol thiozol and imidozol moulding powders including fibre reinforced versions
1 227 756	Courtaulds Ltd.	Polymers reinforced with carbon filaments pre-coated with a thermoplastic polymer deposited from solution
1 232 453	International Synthetic Rubber Co.	Propylene polymer or copolymer moulding compounds containing a hydrogenated rubbery polymer. May be reinforced with talc, asbestos or glass fibres
1 234 208	Courtaulds Ltd.	Monomer casting of caprolactam containing carbon fibres
1 234 844	Phillips Petroleum Co.	Foamable thermoplastic compositions (butadiene/styrene copolymer or polypropylene) with fibrous reinforcement (glass or asbestos fibres) and blowing agent

1 237 844	Monsanto	Polyamides filled or reinforced: fillers treated with various specified coupling agents
1 238 476	BASF AG	Asbestos-filled polyolefin moulding compositions: with a modified adhesion promoter
1 242 124	Farbwerke Hoechst	Polypropylene/asbestos compositions stabilised with various specified stabilisers
1 245 377	Avisun Corp.	Compositions of polypropylene (or copolymer) with 10–80% by weight of chrysotile asbestos and calcium silicate or talc. Natural calcium silicate (Wollastonite) preferred
1 246 034 1 247 794	BASF AG	Polyolefin moulding compositions with 'powdered asbestos' (chrysotile, preferably under 0·1 mm), adhesion promoters and other additives
1 251 641	NRDC	Polymers reinforced with 4–60% by weight of fibres of specified strength and aspect ratio. Fibres may be added to melt solution or precursor of polymer
1 285 828	Celanese Corporation of America	Poly(alkylene terephthalate) moulding compositions. Intimate blends of a reinforcing filler with PET or PBT. Filler (glass fibre) contents of 2–80% mentioned. Lower processing temperatures, mould temperatures and cycle times claimed for PBT in comparison with filled PET
1 285 829	Celanese Corporation of America	Moulding compositions of reduced flammability, comprising an intimate blend of (i) polypropylene terephthalate or PBT; (ii) reinforcing filler (glass fibre 2–60% by weight); (iii) a group V(b) metal (Sn, As, Sb); (iv) an aromatic halogen compound (e.g. tetrabromophthalic or tetrachlorophthalic anhydride or decabromobiphenyl ether). Ratio of the halogen to group V(b) metal between 0·3:1 and 4:1.

APPENDIX 2

ABBREVIATIONS

POLYMERS

ABS	:	Acrylonitrile/butadiene styrene copolymer
CTFE	:	Trifluorochloroethylene polymer
ETFE	:	Ethylene/tetrafluoroethylene copolymer
FEP	:	Fluorinated ethylene/propylene copolymer
HDPE	:	High density polyethylene
LDPE	:	Low density polyethylene
N	:	Nylon
PA	:	Polyacetal
PBT	:	Polybutylene terephthalate
PC	:	Polycarbonate
PET	:	Polyethylene terephthalate
PFA	:	Perfluoroalkoxy polymer
PMMA	:	Polymethyl methacrylate
PP	:	Polypropylene
PPO	:	Modified polyphenylene oxide
PPS	:	Polyphenylene sulphide
PS	:	Polystyrene
PSu	:	Polysulphone
PTFE	:	Polytetrafluoroethylene
PU	:	Polyurethane
PVC	:	Polyvinyl chloride
PVDF	:	Polyvinylidene fluoride
SAN	:	Styrene/acrylonitrile copolymer

REINFORCEMENTS

A or AF	:	Asbestos or asbestos fibre
B	:	Ballotini
CF	:	Carbon fibre
FY	:	Potassium titanate whiskers (Fybex^{R} : DuPont)
GF	:	Glass fibre
T	:	Talc

APPENDIX 3

UNITS: CONVERSION TABLES

The information in these tables has been extracted from *SI Units* compiled by I. G. C. Dryden and published by the British Coal Utilisation Research Association. It is reproduced here by kind permission of the Director General of BCURA.

The abbreviations for multiples and sub-multiples of units, together with the corresponding prefixes, are listed separately below. The use of factors that are not powers of 1000 (shown in italics in the list) is discouraged in the SI system.

Factor	Prefix	Symbol
10^{12}	tera	T
10^{9}	giga	G
10^{6}	mega	M
10^{3}	kilo	k
10^{2}	*hecto*	*h*
10^{1}	*deca*	*da*
10^{-1}	*deci*	*d*
10^{-2}	*centi*	*c*
10^{-3}	milli	m
10^{-6}	micro	μ
10^{-9}	nano	n
10^{-12}	pico	p
10^{-15}	femto	f
10^{-18}	atto	a

BASIC UNITS

SI units					Related units, some permitted in conjunction		Special British and US units		
Length									
m (metre)	km	mm	μm (micrometre or micron)		cm[a]	Å[a] (ångström)	ft		in
1	10^{-3}	10^{3}	10^{6}		10^{2}	10^{10}	3·28		
0·305							1		
		1							$3·94 \times 10^{-2}$
		25·4							1
Mass									
kg (kilogramme)	Mg or t (tonne)	g (gramme)	mg	μg			UK (long) ton	US (short) ton	lb
1	10^{-3}	10^{3}	10^{6}	10^{9}					2·205
0·4536		453·6							1
10^{3}	1						0·984	1·103	2205
	1·016						1	1·12	2240
	0·907						0·893	1	2000
Time									
s (second)	ks	ms	μs	ns	year[b]	day	h (hour)		min (minute)
1	10^{-3}	10^{3}	10^{6}	10^{9}	$3·169 \times 10^{-8}$	$1·157 \times 10^{-5}$	$2·778 \times 10^{-4}$		$1·667 \times 10^{-2}$

Basic units—*continued*

SI units				Related units, some permitted in conjunction			Special British and US units	
Absolute temperature								
°K (degree Kelvin)				°C (degree Celsius)			°F (degree Fahrenheit)	°R (degree Rankine)
y				y − 273·15			32 + 1·8 (y − 273·15)	1·8 y
Electric current								
A (ampere)	kA	mA	μA					
1	10^{-3}	10^{3}	10^{6}					
Luminous intensity								
cd (candela)	Mcd	kcd						
1	10^{-6}	10^{-3}						
Amount of molecular substance								
				kmol	mol	molecules		
				1	10^{3}	602×10^{24}		

[a] Discouraged.
[b] Tropical year for 1900 January 0 at 12 h ephemeris time (11th CGPM, 1960); the SI second is defined so that the tropical year = 31 556 925·9747 s.

DERIVED UNITS

SI units		Related units, some permitted in conjunction		Special British and US units		
Area						
m^2		cm^{2a}		ft^2	in^2	
1		10^4		10·76		
0·093				1		
		1			0·155	
		6·45			1	
Volume						
m^3	mm^3	cm^3	1 (litre) (only if precision < 1 : 10 000)	ft^3	UKgal	USgal
1	10^9	10^6	10^3	35·3		
$2{\cdot}83 \times 10^{-2}$			28·3	1	6·23	
			1		0·22	0·264
			4·546		1	1·201
			3·785		0·833	1
Density						
kg/m^3	Mg/m^3	kg/l	g/cm^{3a}	lb/ft^3	lb/gal	
1		10^{-3}	10^{-3}	$6{\cdot}24 \times 10^{-2}$	$1{\cdot}002 \times 10^{-2}$	
10^3	1	1	1	62·4	10(·02)	
16		$1{\cdot}6 \times 10^{-2}$	$1{\cdot}6 \times 10^{-2}$	1		
Mass flux						
$kg/s\ m^2$		$kg/h\ m^2$		$lb/ft^2\ h$		
1		3600		738		
$2{\cdot}778 \times 10^{-4}$		1		0·205		
		4·88		1		

Derived units—*continued*

SI units				Related units, some permitted in conjunction					Special British and US units		
Force											
N (newton)	kg m/s²			kgf[b] (kilo-gramme-force)	Mdyn[b] (megadyne)	dyn[b]			lbf		
1	1				10^{-1}	10^5					
9·807				1	0·981				2·205		
				0·0454	0·445				1		
10				1·02	1	10^6			2·248		
Pressure, stress											
N/m² or Pa (pascal)	kg/s² m	N/mm²		bar	kgf/cm²[b] (at)	dyn/cm²[b]	mmHg[b] (torr)		atm	psi (lbf/in²)	in w.g.
10^5	10^5	0·1		1		10^6	750		0·987	14·5	
				1·013			760		1	14·7	
				0·9807	1		736		0·968	14·2	
$6{\cdot}895 \times 10^3$					$7{\cdot}03 \times 10^{-2}$					1	
249											1
133							1				
Mechanical and electrical energy and heat											
J = N m (joule)	kg m²/s²	MJ	aJ	erg[b] (dyn cm)	kWh	kcal[b]	cal[b]	eV (electron volt)	ft lbf	Btu	therm
1	1	10^{-6}	10^{18}	10^7			0·239		0·738		
		3·6			1	860					
4·187						10^{-3}	1				
						0·252				1	
		105·5			29·3	$2{\cdot}52 \times 10^4$				10^5	1
			0·160 2					1			

Derived units—*continued*

SI units					Related units, some permitted in conjunction	Special British and US units	
Power							
W	J/s	kg m^2/s^3	MW	kW	kcal/h[b]	hp	Btu/h
1	1	1	10^{-6}	10^{-3}	0·86	$1{\cdot}341 \times 10^{-3}$	3·41
0·293							1
Heat flux							
W/m^2		J/s m^2	kg/s^3		kcal/h m^2[b]	Btu/ft^2 h	
1		1	1		0·860	0·317	
1·163		1·163			1	0·369	
3·155		3·155			2·71	1	
Heat transfer coefficient							
W/m^2 degK		J/s m^2 degK	kg/s^3 degK		kcal/h m^2 degC (or degK)[b]	Btu/ft^2 h degF	
1		1	1		0·860	0·176	
1·163		1·163			1	0·205	
5·675		5·675			4·88	1	
Thermal conductivity							
W/m degK		J/s m degK	kg m/s^3 degK		kcal/h m degC (or degK)[b]	Btu/ft h degF	
1		1	1		0·860	0·578	
1·163		1·163			1	0·672	
1·731		1·731			1·488	1	
Specific heat capacity							
J/kg degK		m^2/s^2 degK	kJ/kg degK		cal/g degC[b]	Btu/lb degF	
1		1	10^{-3}		$2{\cdot}39 \times 10^{-4}$	$2{\cdot}39 \times 10^{-4}$	
$4{\cdot}187 \times 10^3$			4·187		1	1	

Derived units—*continued*

SI units				Related units, some permitted in conjunction		Special British and US units
Dynamic viscosity						
N s/m^2	kg/s m	g/s m	mN s/m^2	P[a] (poise)	cP[a]	lb/ft h
1	1	10^3	10^3	10	10^3	$2{\cdot}42 \times 10^3$
		0·413	0·413		0·413	1
Kinematic viscosity						
m^2/s		mm^2/s		St[a] (stoke)	cSt[a]	ft^2/h
1		10^6		10^4	10^6	$3{\cdot}88 \times 10^4$
		25·8			25·8	1
Surface tension, surface energy						
N/m	kg/s^2	mN/m	J/m^2	dyn/cm[b]	erg/cm^2[b]	
1	1	10^3	1	10^3	10^3	
Electrical conductivity						
S/m (siemen/m)	Ω^{-1} m^{-1}	A^2 s^3/kg m^3		mho/m	mho/cm[a]	
1	1	1		1	10^{-2}	
Electric field strength						
V/m	W/A m	kg m/A s^3				
1	1	1				
Electrical capacitance						
F (farad)	C/V	A^2 s^4/kg m^2				
1	1	1				

Derived units—*continued*

SI units				Related units, some permitted in conjunction	Special British and US units
Electric (self and mutual) inductance					
H (henry)	V s/A	$kg\ m^2/A^2\ s^2$			
1	1	1			
Electric charge					
C (coulomb)	A s				
1	1				
Electric potential					
V (volt)	W/A	$kg\ m^2/A\ s^3$			
1	1	1			
Electrical resistance					
Ω (ohm)	V/A	W/A^2	$kg\ m^2/A^2\ s^3$		
1	1	1	1		
Periodic frequency					
Hz ($\equiv s^{-1}$) (hertz)					
Rotational frequency					
s^{-1}					

INDEX